KS3 Success

Workbook

Mathematics

Fiona C. Mapp

Contents

Number

Algebra

Shape, space and measures

Handling data

Homework diary

TOPIC	STUDY DATE	SCORE
Numbers		/30
Positive & negative numbers		/32
Working with numbers		/29
Fractions		/33
Decimals		/36
Percentages 1		/31
Percentages 2		/32
Equivalents & using a calculator		/33
Checking calculations		/35
Ratio		/27
Indices		/40
Standard index form		/36
Algebra 1		/33
Algebra 2		/33
Equations 1		/35
Equations 2 & inequalities		/36
Patterns & sequences		/25
Coordinates & graphs		/17
More graphs		/26
Interpreting graphs		/16
Shapes		/28
Solids		/18
Constructions and LOGO		/18
Loci & coordinates in 3D		/17
Angles & tessellations		/24
Bearings & scale drawings		/24
Transformations		/18
Pythagoras' theorem		/25
Trigonometry		/28
Measures & measurement		/32
Similarity		/23
Area & perimeter of 2D shapes		/26
Volume of 3D solids		/23
Collecting data		/17
Representing information		/13
Scatter diagrams		/16
Averages 1		/21
Averages 2		/21
Cumulative frequency		/17
Probability 1		/19
Probability 2		/17

Progress plotter

Once you have completed a Topic, and filled in your score on the Homework Diary opposite, use this Progress plotter to chart your success! Fill in the boxes with your score for each unit and watch your results get better and better.

	Nearly all right – Excellent work!	More than half – Good but keep trying.	Less than half – Room for improvement.	Under 5 – Needs more work.
Numbers				
Positive & negative numbers				
Working with numbers				
Fractions				
Decimals				
Percentages 1				
Percentages 2				
Equivalents & using a calculator				
Checking calculations				
Ratio				
Indices				
Standard index form				
Algebra 1				
Algebra 2				
Equations 1				
Equations 2 & inequalities				
Patterns & sequences				
Coordinates & graphs				
More graphs				
Interpreting graphs				
Shapes				
Solids				
Constructions and LOGO				
Loci & coordinates in 3D				
Angles & tessellations				
Bearings & scale drawings				
Transformations				
Pythagoras' theorem				
Trigonometry				
Measures & measurement				
Similarity				
Area & perimeter of 2D shapes				
Volume of 3D solids				
Collecting data				
Representing information				
Scatter diagrams				
Averages 1				
Averages 2				
Cumulative frequency				
Probability 1				
Probability 2				

Numbers

A

Choose just one answer, a, b, c or d.

1 From this list of numbers: 4, 7, 9, 20 which one is a prime number? *(1 mark)*

a) 4 ☐ b) 7 ☑ c) 9 ☐ d) 20 ☐

2 The positive square root of 64 is: *(1 mark)*

a) 7 ☐ b) –8 ☐ c) –7 ☐ d) 8 ☑

3 Work out the lowest common multiple of 6 and 8. *(1 mark)*

a) 48 ☐ b) 3 ☐ c) 24 ☑ d) 2 ☐

4 What is the value of 3^3? *(1 mark)*

a) 9 ☐ b) 12 ☐ c) 72 ☐ d) 27 ☑

5 What is the reciprocal of $\frac{7}{9}$? *(1 mark)*

a) $\frac{9}{7}$ ☑ b) $\frac{7}{9}$ ☐ c) $\frac{9}{5}$ ☐ d) 9 ☐

6 Calculate the highest common factor of 18 and 24. *(1 mark)*

a) 6 ☑ b) 18 ☐ c) 12 ☐ d) 432 ☐

Score   / 6

B

Answer all parts of all questions.

1 Look at the numbers in the cloud.

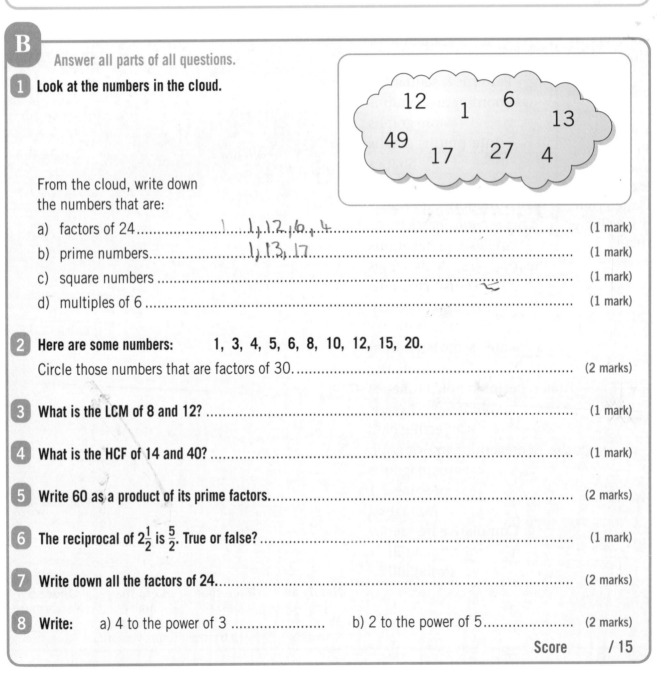

12   1   6   13
49   17   27   4

From the cloud, write down the numbers that are:

a) factors of 24 .... 1, 12, 6, 4 .... *(1 mark)*

b) prime numbers .... 1, 13, 17 .... *(1 mark)*

c) square numbers .... *(1 mark)*

d) multiples of 6 .... *(1 mark)*

2 Here are some numbers: 1, 3, 4, 5, 6, 8, 10, 12, 15, 20.

Circle those numbers that are factors of 30. .... *(2 marks)*

3 What is the LCM of 8 and 12? .... *(1 mark)*

4 What is the HCF of 14 and 40? .... *(1 mark)*

5 Write 60 as a product of its prime factors. .... *(2 marks)*

6 The reciprocal of $2\frac{1}{2}$ is $\frac{5}{2}$. True or false? .... *(1 mark)*

7 Write down all the factors of 24. .... *(2 marks)*

8 Write: a) 4 to the power of 3 ....   b) 2 to the power of 5 .... *(2 marks)*

Score   / 15

Answer all parts of the questions.

1 **William has 12 counters. He can put them in 3 rows, with 4 counters in each row.**

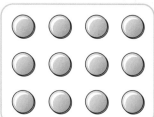

a) Draw a diagram to show how he can put 12 counters in 2 rows, with the same number of counters in each row.

000000
0 00000

(1 mark)

b) Draw a diagram to show a different way that William can put 12 counters in a different number of rows, with the same number of counters in each row.

000
000
000
000

(1 mark)

c) Fill in the table to show how many rows William can make with 12 counters, and how many counters there are in each row.

Number of rows	Number of counters in each row
1 row	12 counters in each row
...2... rows	...6... counters in each row
3 rows	...4... counters in each row
...4... rows	...3.... counters in each row
...6... rows	...2... counters in each row
...12... rows	1 counter in each row

(4 marks)

d) William says: *I can put 12 counters in 5 rows with the same number of counters in each row.*

Explain why William is wrong. .....Because.....5.....is not a multiple of 12.... (1 mark)

2 **Penelope and Tom are describing some numbers. Decide which number each is thinking about.**

Penelope   *My number is a square number, greater than 10 and less than 20.*

Tom   *My number is the only even prime number.*

Penelope's number is ....16.....................

Tom's number is ....2.....................

(2 marks)

Score   / 9

Total score   / 30

How well did you do?   ✗ 1–8 Try again   9–15 Getting there   16–24 Good work   25–30 Excellent! ✓

For more help on this topic see KS3 Maths 5–8 Success Guide pages 4–5.

7

NUMBERS Number

Positive & negative numbers

A

Answer all parts of all questions.

1 Here are some number cards:

$\boxed{5}\ \boxed{-3}\ \boxed{2}\ \boxed{-7}\ \boxed{0}$

a) Choose two cards that add up to give –10. ....–3 –7.......... (1 mark)

b) Choose two cards that add up to give –7. ......0 – 7...... (1 mark)

c) Choose two cards that multiply to give –15. ......5 – 3...... (1 mark)

2 Each of the calculations below has an answer that is one of the numbers from the list:

12, –12, –3, –4, 3, 8.

Select the correct number for each answer.

a) –3 × –4 = ...–12......

b) –16 ÷ (–2) = ...8......

c) 8 – (–4) = ...–4......

d) 4 + (–7) = ...–3......

e) –9 ÷ (–3) = ...–3......

f) –7 – (–3) = ...4...... (6 marks)

3 Work out the following.

a) –7 × –3 ......–21......

b) –9 – (–4) ......5......

c) 12 ÷ (–6) ......–2......

d) –8 × 2 ......–16...... (4 marks)

4 The temperature at midday was –12°C. By midnight it had dropped by 10 degrees.

What was the temperature at midnight?

......–22°C...... (1 mark)

5 Here are some numbers in a number pyramid. The number in each rectangle is found by adding the two numbers below.

Complete the number pyramid.

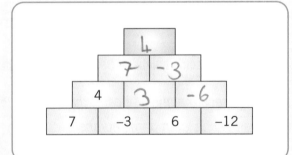

(2 marks)

6 Here are some signs:

$\boxed{+}\ \boxed{\times}\ \boxed{\div}\ \boxed{/}$

Insert the correct sign to make each calculation correct.

a) –12 ......÷...... –3 = 4

b) –7 ......–...... –3 = –4

c) 9 ......–...... –3 = –3

d) 12 ......+...... –5 = 7 (4 marks)

Score / 20

B

Answer all parts of the questions.

1 The arrow on the thermometer shows a temperature of 20°C.

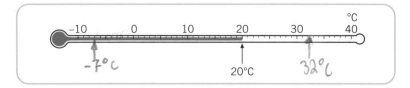

-7°c   20°C   32°C

a) Draw an arrow on the thermometer to show a temperature of 32°C.
   Label the arrow 32°C.   (1 mark)

b) Draw an arrow on the thermometer to show a temperature of –7°C.   (2 marks)

c) The temperature was 25°C. It fell by 27 degrees.

   What is the temperature now?......–2°c..   (1 mark)

d) The temperature was –9°C. It rose by 21 degrees.

   What is the temperature now? ....12°C...   (1 mark)

e) Write these temperatures in order, highest first.

   –6°C, 9°C, –3°C, 15°C, 0°C, –12°C

   ....15.......°C, ...9.......°C, ...0........°C, ...–3.......°C, ...–6......°C, ...–12.....°C
   highest   lowest

                                                                                        (1 mark)

2 Here is a list of numbers:

   –9, –5, –1, 0, 3, 2, 5

Choose a number from the list to make each statement correct.

a) 3 – ....–1..... = 4   b) 5 × ....–5.... = –25

c) –9 + ...5...... = –4   d) 2 – ....–5.... = 7

e) –9 ÷ .....3....... = –3   (5 marks)

f) What is the total of all seven numbers in the list?   ....–5.................. (1 mark)

Score   **/ 12**

Total score   **/ 32**

How well did you do?   ✗ 1–9 **Try again**   10–17 **Getting there**   18–25 **Good work**   26–32 **Excellent!** ✓

For more help on this topic see KS3 Maths 5–8 Success Guide pages 6–7.

Working with numbers

A Answer all parts of all questions.

1 You have four tins of beans with the weights written on them.

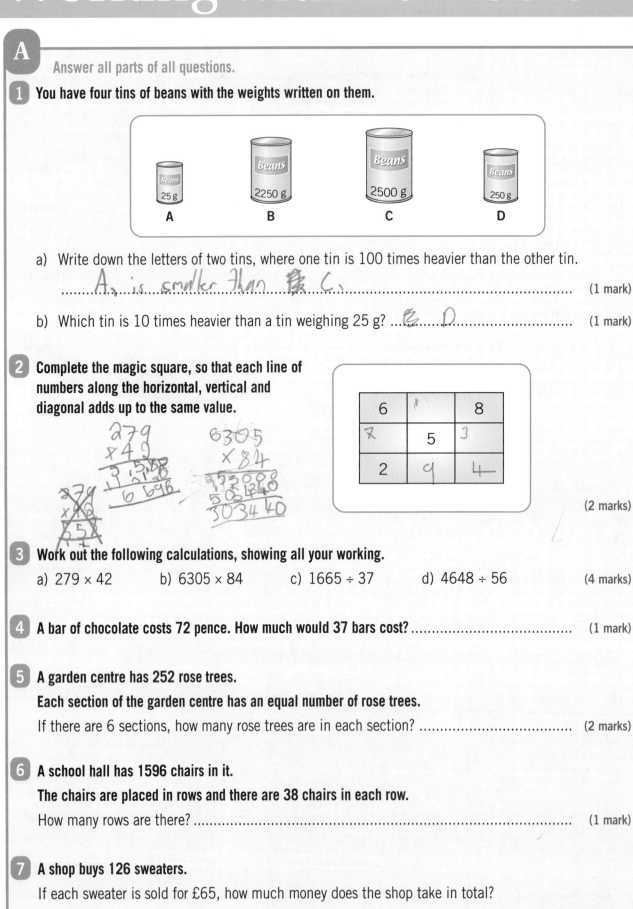

25 g	2250 g	2500 g	250 g
A	B	C	D

a) Write down the letters of two tins, where one tin is 100 times heavier than the other tin.

...........A, is smaller than B C............................. (1 mark)

b) Which tin is 10 times heavier than a tin weighing 25 g? ....B...D.......... (1 mark)

2 Complete the magic square, so that each line of numbers along the horizontal, vertical and diagonal adds up to the same value.

279
× 42
...

6305
× 84
...

6		8
7	5	3
2	9	4

(2 marks)

3 Work out the following calculations, showing all your working.

a) 279 × 42   b) 6305 × 84   c) 1665 ÷ 37   d) 4648 ÷ 56   (4 marks)

4 A bar of chocolate costs 72 pence. How much would 37 bars cost? .... (1 mark)

5 A garden centre has 252 rose trees.

Each section of the garden centre has an equal number of rose trees.

If there are 6 sections, how many rose trees are in each section? .... (2 marks)

6 A school hall has 1596 chairs in it.

The chairs are placed in rows and there are 38 chairs in each row.

How many rows are there? .... (1 mark)

7 A shop buys 126 sweaters.

If each sweater is sold for £65, how much money does the shop take in total?

.... (2 marks)

Score   / 14

B

Answer all parts of the questions.

Handwritten at top:
$3 \times 4 = 12$ $1200 \div ? = 60$
$1200 = 60 \times ?$
$3 = 12 \div 4$
$4 = 12 \div 3$
$? = 1200 \div 60$

1 There are forty-five pencils in a box.
A box of pencils costs £3.50.

a) How many pencils are there in 6 boxes? .... pencils (1 mark)

b) How much do 7 boxes cost? £ .... (1 mark)

c) How much do 135 pencils cost? £ .... (1 mark)

d) How many boxes of pencils can be bought with £42? .... boxes (1 mark)

2 Fill in the missing numbers so
that the answer is always 60.

Handwritten:
200
$\times 60$
1200

220
$\times 60$
1200

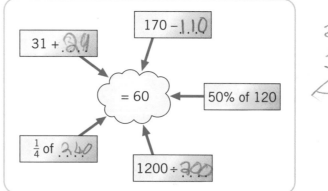

31 + .29

170 – 1.10

= 60

50% of 120

¼ of 2.40

1200 ÷ 2.0

(4 marks)

3 Nigel owns a garden centre. He sells small shrubs for £6.75 each.

a) Nigel sells 37 small shrubs one weekend to Mrs Robinson.
How much does he get for the 37 shrubs? £ .... (2 marks)

b) Each small shrub needs 60 grams of compost when it is planted.
Mrs Robinson buys a sack of compost weighing 2000 grams.

i) How many shrubs can be planted? .... shrubs (2 marks)

ii) How much more compost does she need in order to
plant all 37 shrubs? .... grams (1 mark)

4 Raj puts a three-digit whole number into his calculator.
He multiplies it by 10.

a) Fill in one other digit that you know
must be on the calculator display.

(1 mark)

b) Raj starts again with the same three-digit
number. This time he multiplies it by 1000.
Fill in all the digits that might be on the
calculator display.

(1 mark)

Score / 15

Total score / 29

How well did you do?   ✗ 1–8 Try again   9–17 Getting there   18–24 Good work   25–29 Excellent! ✓

For more help on this topic see KS3 Maths 5–8 Success Guide pages 8–9.

Fractions

A

Choose just one answer, a, b, c or d.

1 In a class of 32 students, $\frac{3}{4}$ are right-handed. How many students are right-handed? (1 mark)

a) 8 ☐   b) 16 ☐   c) 26 ☐   d) 24 ☐

2 Work out the answer to $\frac{5}{7} - \frac{1}{21}$. (1 mark)

a) $\frac{4}{14}$ ☐   b) $\frac{14}{21}$ ☐   c) $\frac{2}{7}$ ☐   d) $\frac{1}{3}$ ☐

3 Work out the answer to $\frac{3}{5} \div \frac{4}{10}$. (1 mark)

a) $\frac{12}{60}$ ☐   b) $\frac{3}{2}$ ☐   c) $\frac{7}{15}$ ☐   d) $\frac{12}{15}$ ☐

4 Work out the answer to $\frac{2}{9} \times \frac{3}{11}$. (1 mark)

a) $\frac{22}{27}$ ☐   b) $\frac{6}{9}$ ☐   c) $\frac{5}{20}$ ☐   d) $\frac{6}{99}$ ☐

Score   / 4

B

Answer all parts of all questions.

1 Work out the following:

a) $\frac{2}{7} + \frac{3}{5} =$ ....

b) $\frac{8}{13} - \frac{1}{2} =$ .... (2 marks)

2 Arrange these fractions in order of size, smallest first.

a) $\frac{4}{5}, \frac{1}{3}, \frac{2}{9}, \frac{4}{7}, \frac{1}{2}$ ....

b) $\frac{9}{13}, \frac{3}{5}, \frac{2}{7}, \frac{1}{3}$ .... (2 marks)

3 Work out the following:

a) $\frac{2}{7} \times \frac{4}{9}$ ....

b) $\frac{6}{7} \div \frac{3}{14}$ ....

c) $\frac{9}{11} \times \frac{2}{5}$ ....

d) $\frac{4}{5} \div \frac{3}{10}$ .... (4 marks)

4 Work out the amounts in each statement and decide whether the statement is true or false.

a) $\frac{3}{5}$ of 20 is greater than $\frac{2}{7}$ of 14.   .... (1 mark)

b) $\frac{9}{11}$ of 44 is smaller than $\frac{1}{3}$ of 72.   .... (1 mark)

c) $\frac{2}{7}$ of 21 is greater than $\frac{3}{5}$ of 15.   .... (1 mark)

d) $\frac{1}{3}$ of 27 is smaller than $\frac{5}{9}$ of 45.   .... (1 mark)

5 The village of Grimley has 2200 inhabitants. $\frac{3}{5}$ of the inhabitants are men.

How many men are in the village? .... (1 mark)

6 Mark has a fruit stall. $\frac{2}{7}$ of the fruit he has are apples.

If he has 245 pieces of fruit in total, how many are apples? .... (1 mark)

7 A hospital has 350 pints of milk delivered. $\frac{2}{5}$ of the milk is skimmed.

How many pints are not skimmed? .... (1 mark)

Score   / 15

Answer all parts of the questions.

1 **Rani and David bake some cakes. They have 16 cakes each.**

a) Rani eats a quarter of her cakes.
How many cakes does Rani eat? .... (1 mark)

b) David eats 12 of his 16 cakes.
What fraction of his cakes does David eat? .... (1 mark)

c) How many cakes are left altogether? .... (1 mark)

2 **Fill in the missing values so that the answer is always 8.**

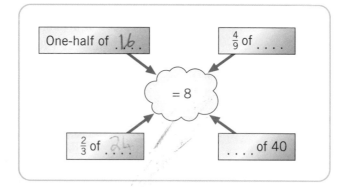

One-half of .16.

$\frac{4}{9}$ of . . . .

= 8

$\frac{2}{3}$ of .24.

. . . . of 40

(4 marks)

3 **Carry out these calculations.**

a) $\frac{1}{2}$ of £60   ....£30............ (1 mark)

b) $\frac{2}{3}$ of £900   £300
.... (1 mark)

c) $\frac{2}{7} + \frac{3}{5}$ .... (1 mark)

d) $\frac{4}{7} \times \frac{2}{5}$ .... (1 mark)

4 **In a book there are three pictures on the same page.**

Picture 1 uses $\frac{1}{4}$ of the page.

Picture 2 uses $\frac{1}{16}$ of the page.

Picture 3 uses $\frac{1}{64}$ of the page.

a) In total, what fraction of the page do pictures 1 and 2 make up? .... (2 marks)

b) To put a picture in a book
is more expensive than text.

Cost of picture = £35 for every $\frac{1}{8}$ of a page

If a picture takes up $\frac{3}{32}$ of a page, how much will it cost? .... (1 mark)

Score   / 14

Total score   / 33

How well did you do?   ✗ 1–12 Try again   13–22 Getting there   23–28 Good work   29–33 Excellent! ✓

FRACTIONS   Number

For more help on this topic see KS3 Maths 5–8 Success Guide pages 10–11.

13

Decimals

A

Choose just one answer, a, b, c or d.

1 **Work out the answer to 9.23 × 6.** (1 mark)

a) 55.83 ☐ b) 54.83 ☐
c) 54.38 ☐ d) 55.38 ☐

2 **Round 17.26 to one decimal place.** (1 mark)

a) 17.3 ☐ b) 17.2 ☐
c) 1.726 ☐ d) 1.73 ☐

3 **Round 14.635 to two decimal places.** (1 mark)

a) 14.63 ☐ b) 14.65 ☐
c) 146.35 ☐ d) 14.64 ☐

4 **If a piece of cake weighs 0.2 kg, how much would 60 identical pieces of cake weigh?** (1 mark)

a) 0.12 kg ☐ b) 1.2 kg ☐
c) 12 kg ☐ d) 120 kg ☐

5 **A piece of writing paper is 0.01 cm thick. A notepad has 120 sheets of paper. How thick is the notepad?** (1 mark)

a) 1.2 cm ☐ b) 0.12 cm ☐
c) 12 cm ☐ d) 120 cm ☐

Score   / 5

B

Answer all parts of all questions.

1 **Here are some numbers:**

4.39   3.69   3.691   2.71   4.385   3.62   4.38

Arrange these numbers in order of size, smallest first.

.... (2 marks)

2 **The number pyramid right is found by adding the two numbers in the row below.**

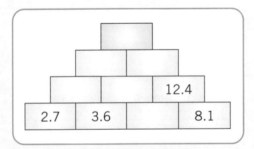

Complete this pyramid.

(4 marks)

3 **Four friends are playing a game. Their scores are shown below:**

Tom	Ahmed	Rebecca	Lin
4.27	4.98	5.07	5.065

a) Who won the game, with the highest score? .... (1 mark)

b) What is the difference between Ahmed's and Lin's scores? .... (1 mark)

14

4 A CD costs £11.58. James buys six. How much must he pay for the six CDs?

.... (1 mark)

5 For each statement decide whether it is true or false.

a) 2.73 rounded to 1 decimal place is 2.7 .... (1 mark)

b) 12.738 rounded to 2 decimal places is 12.73 .... (1 mark)

c) 25.621 rounded to 1 decimal place is 25.2 .... (1 mark)

6 Here are some number cards:

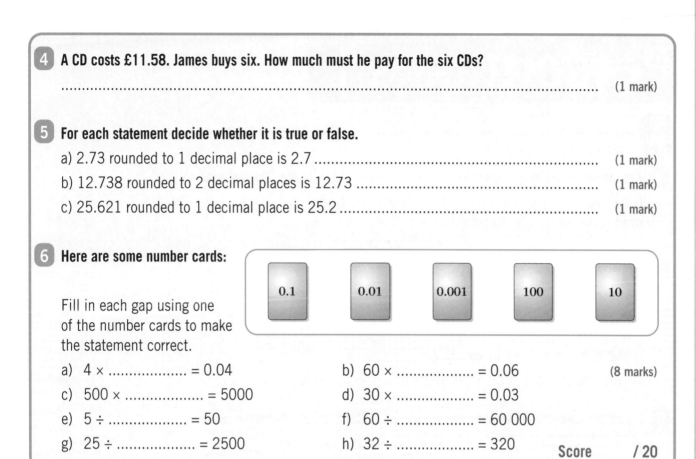

Fill in each gap using one of the number cards to make the statement correct.

a) $4 \times$ .... $= 0.04$ b) $60 \times$ .... $= 0.06$ (8 marks)

c) $500 \times$ .... $= 5000$ d) $30 \times$ .... $= 0.03$

e) $5 \div$ .... $= 50$ f) $60 \div$ .... $= 60\,000$

g) $25 \div$ .... $= 2500$ h) $32 \div$ .... $= 320$

Score   / 20

C

Answer all parts of the questions.

1 Complete these calculations to make them correct.

a) $14.6 + 3.9 = 12.1 +$ .... (1 mark)

b) $8.3 \times 4 = 2 \times$ .... (1 mark)

c) $937.65 - 2.3 = 654.1 +$ .... (1 mark)

d) $124.8 \div 4 = 2 \times$ .... (1 mark)

2 Look at these number cards:

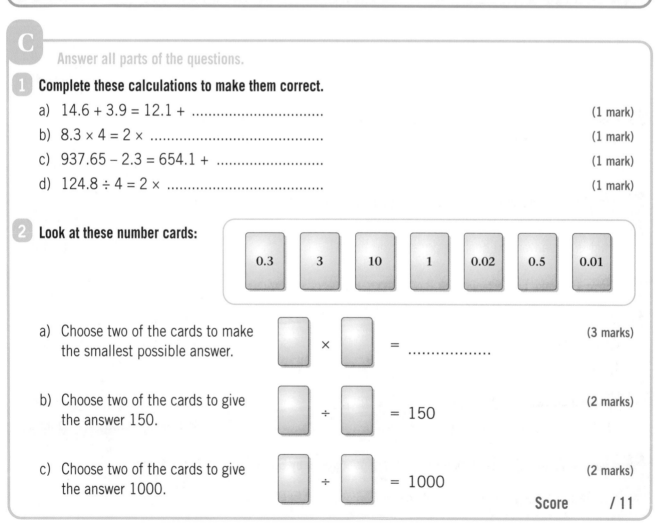

a) Choose two of the cards to make the smallest possible answer.

□ × □ = ... (3 marks)

b) Choose two of the cards to give the answer 150.

□ ÷ □ = 150 (2 marks)

c) Choose two of the cards to give the answer 1000.

□ ÷ □ = 1000 (2 marks)

Score   / 11

Total score   / 36

How well did you do?   ✗ 1–12 Try again   13–24 Getting there   25–29 Good work   30–36 Excellent! ✓

For more help on this topic see KS3 Maths 5–8 Success Guide pages 12–13.

15

DECIMALS   Number

Percentages 1

A

Answer all parts of all questions.

1 The building society has reported a 10% rise in house prices over the last year.

If the average price of a house last year was £160 000,
what is the average house price this year? .... (2 marks)

2 A shoe shop is having a sale:

Shoe World
15% off
marked prices

A pair of trainers originally cost £130.
What is the sale price of the trainers? .... (2 marks)

3 Match the calculations with the correct answers. The first has been done for you.

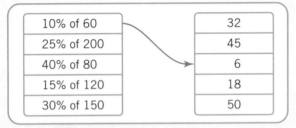

10% of 60	32
25% of 200	45
40% of 80	6
15% of 120	18
30% of 150	50

(4 marks)

4 Police reported an increase in burglaries last year of 15%. If 620 burglaries had been committed the year before, how many more burglaries had been committed?

.... (1 mark)

5 A train ticket from Manchester to Birmingham costs £24 return. If you book in advance,
ⓒ the price of the ticket is reduced by 6%.
What is the advanced booking price of the ticket? .... (1 mark)

6 These are some of the results for Hywel's end of year examinations:
ⓒ

Maths	64 out of 80	= ....%	English	27 out of 50	= ....%
Geography	39 out of 42	= ....%	Science	121 out of 150	= ....%
German	42 out of 110	= ....%			

a) Change each of the results into a percentage. (2 marks)
b) In which subject did Hywel do worst? .... (1 mark)

7 A survey showed the favourite flavours of crisps, chosen by 500 people were as follows:
ⓒ

Salt 'n' vinegar 204   Cheese 196   Beef 100

a) What percentage of people preferred cheese flavoured crisps? .... (2 marks)
b) What percentage of people did not choose beef flavoured crisps? .... (2 marks)

Score   / 17

 Indicates that a calculator may be used

Answer all parts of the questions.

1 **Complete the missing values.**

10% of £80 = £.. (1 mark)

5% of £80 = £.. (1 mark)

$2\frac{1}{2}$% of £80 = £.. (1 mark)

$17\frac{1}{2}$% of £80 = £.. (1 mark)

2 **A cake shop sells some different types of cakes.**

C **The table shows how many they sold in one day.**

Cake	Number of cakes sold	Takings (£)
Doughnut	146	32.12
Battenberg	45	8.10
Coconut slice	37	20.35
Almond slice	48	41.76
Chocolate brownie	74	70.30
Total	350	172.63

a) What percentage of the number of cakes sold were doughnuts?

Show your working..% (2 marks)

b) What percentage of the total takings was for doughnuts?

Show your working..% (2 marks)

c) What percentage of the number of cakes sold were Battenberg?

Show your working..% (2 marks)

3 **Calculate:**

C a) i) 4% of £32.50 = £.. (1 mark)

ii) $12\frac{1}{2}$% of £78 = £.. (1 mark)

b) A television costs £199. In a sale, 28% is taken off the price.
Work out the sale price of the television set.

£.. (2 marks)

Score   / 14

Total score   / 31

How well did you do?   ✗ 1–11 **Try again**   12–17 **Getting there**   18–25 **Good work**   26–31 **Excellent!** ✓

For more help on this topic see KS3 Maths 5–8 Success Guide pages 14–15.

17

Percentages 2

A

Choose just one answer, a, b, c or d.

1 A shop bought a camera for £50.
A customer later buys the camera for £65.
Find the percentage profit. (1 mark)

a) 20% ☐   b) 15% ☐

c) 30% ☐   d) 40% ☐

2 VAT at 20% is added to a telephone bill of £80.
What is the total amount to be paid? (1 mark)

a) £90 ☐   b) £95 ☐

c) £85 ☐   d) £96 ☐

3 A flat is bought for £80 000.
Each year the price rises by 10%.
How much is the flat worth after
2 years? (1 mark)

a) £96 000 ☐   b) £98 000 ☐

c) £96 500 ☐   d) £96 800 ☐

4 LEVEL 8 A sweater is bought in a sale for £20
after a 15% reduction. What was the original
price of the sweater? (1 mark)

a) £25 ☐

b) £23.53 ☐

c) £23.55 ☐

d) £23 ☐

5 LEVEL 8 The price of a meal after VAT at 20%
has been added is £102.24.
What was the price of the meal before VAT?
(1 mark)

a) £82.50 ☐

b) £87.50 ☐

c) £85.11 ☐

d) £85.20 ☐

Score   / 5

B

Answer all parts of all questions.

1 John bought a sports car for £18 500. Two years later he sold the sports car for £14 000.
Work out the percentage loss.

.... (2 marks)

2 On April 1st a train journey increases in price by 5%. Six months later it increases in price again by 2%.
Work out the final price of a ticket that originally cost £40.

.... (2 marks)

3 A car was bought for £12 000. Each year it depreciated in value by 5%. Work out the value of the
car after 3 years.

.... (2 marks)

4 LEVEL 8 After a 7% rise, Frances is earning £390.55 per week. What was she earning before?

.... (2 marks)

5 LEVEL 8 The Patel family paid £4525 for a holiday to New Zealand, after a 10% surcharge was
added at the last minute. What did they originally think they would be paying?

.... (2 marks)

6 LEVEL 8 An electricity bill is £85 after a 5% reduction is made. How much was it originally?

.... (2 marks)

Score   / 12

Ⓒ *Indicates that a calculator may be used*

Answer all parts of the questions.

1 The table shows some information about pupils in a school:

	Wear glasses	Do not wear glasses
Girls	6	120
Boys	14	110

There are 250 pupils in this sample.

a) What percentage of the pupils are girls?

Show your working... % (2 marks)

b) The following month 2 more of the boys and 3 more of the girls have to wear glasses.
Work out the percentage increase of those pupils who now wear glasses.
(There are still 250 pupils in the sample.)

Show your working... % (2 marks)

2 A furniture shop had a closing down sale. The sale started on Wednesday and finished on Sunday.
For each day of the sale, the prices were reduced by 20% of the prices of the day before.

a) A sofa had a price of £850 on Tuesday and it was bought on Friday. How much was it bought for?

Show your working... (3 marks)

b) LEVEL 8 A sofa is bought on Thursday for £650. What was the price on Wednesday?

Show your working... (2 marks)

c) LEVEL 8 Another shop is reducing its prices by 15% each day. How many days would it take for
its original prices to be reduced by more than 50%?

Show your working... days (2 marks)

3 a) **Roy sells used cars. He bought a car for £3570.**
Two days later he sold it for £5250. Work out his percentage profit.

Show your working... % (2 marks)

b) LEVEL 8 **Roy sells a car for £6800. He makes a 30% profit.**

For what price did Roy buy the car? £ .... (2 marks)

Score   / 15

Total score   / 32

How well did you do? ✗ 1–8 **Try again** 9–15 **Getting there**   16–24 **Good work** 25–32 **Excellent!** ✓

For more help on this topic see KS3 Maths 5–8 Success Guide pages 16–17.

19

PERCENTAGES 2   Number

Equivalents & using a calculator

A

Answer all parts of all questions.

1 Change these fractions into   i) decimals and   ii) percentages:

a) Fraction = $\frac{3}{5}$ i) decimal = ...   ii) percentage = ...%   (2 marks)

b) Fraction = $\frac{1}{100}$ i) decimal = ...   ii) percentage = ...%   (2 marks)

c) Fraction = $\frac{1}{8}$ i) decimal = ...   ii) percentage = ...%   (2 marks)

2 Here are some cards. Divide the cards into four sets of three cards each showing the same value.

0.3 $\frac{9}{20}$ 0.1 45% 25% $\frac{3}{10}$ 0.45 $\frac{1}{4}$ 30% 10% 0.25 $\frac{1}{10}$

Set 1   ...., ..., ...   (1 mark)

Set 2   ...., ..., ...   (1 mark)

Set 3   ...., ..., ...   (1 mark)

Set 4   ...., ..., ...   (1 mark)

3 Here are some cards:

$\frac{2}{5}$ 72% $\frac{3}{8}$ 25% 0.9 0.41 30%

Put these cards in order of size, smallest first .... (2 marks)

4 **C** The same television is being sold in different shops.

SUPERS	Electricals	RAMONES
TV £570	TV £510	TV £375
NOW $\frac{1}{3}$ off	NOW 20% off	plus VAT at 20%

Work out the price of the television in each of the three shops.

Which shop sells the television at the lowest price? .... (4 marks)

5 Work out these calculations without using a calculator.

a) $5 + 3 \times 2 =$ ....

b) $(6 + 3) \times 4 =$ ....

c) $12 + 6 \div 2 =$ .... (3 marks)

6 **C** Use a calculator to work out these calculations.

a) $\dfrac{6.2 + 4.9}{3.7 \times 4.1} =$ ....

b) $\dfrac{\sqrt{10.3} - 2.1}{(2.3)^2} =$ ....

c) $\dfrac{9.3^2 \times 2}{\sqrt{5.6}} =$ .... (3 marks)

Score   / 22

C   *Indicates that a calculator may be used*

B Answer all parts of the questions.

1 **Charlotte has some cards:**

| 0.125 | 0.75 | $\frac{1}{2}$ | 0.4 | 40% | 50% |

Reece has some different cards:

| $\frac{2}{5}$ | 0.5 | 75% | $\frac{1}{8}$ | $\frac{15}{20}$ | 12.5% |

They decide to group the cards together, so that in each group the cards are equivalent.

Complete the card groups below. (4 marks)

$\frac{2}{5}$, ..., ..., 0.5, ..., ..., 12.5%   ...., 0.75, ...

2 **A target board has some fractions, decimals and percentages.**

Here are some of the sets of numbers that are equivalent:

Set 1   55%,   0.55, $\frac{11}{20}$

Set 2   20%,   0.2, $\frac{2}{10}$

0.55	$\frac{1}{25}$	$\frac{1}{3}$	$\frac{11}{20}$	$\frac{2}{7}$
$\frac{3}{8}$	20%	0.9	0.375	$\frac{1}{20}$
90%	$0.\dot{3}$	0.05	$\frac{2}{10}$	$\frac{9}{10}$
0.2	5%	37.5%	$33.\dot{3}\%$	55%

There are four other sets of numbers that are equivalent. Write them in the spaces provided.

Set 3   ...., ..., ... (1 mark)

Set 4   ...., ..., ... (1 mark)

Set 5   ...., ..., ... (1 mark)

Set 6   ...., ..., ... (1 mark)

3 **Work out the values of the following. Give your answers to 1 decimal place.**

C

a) $5.2 + 3 + \dfrac{4 \times \sqrt{3.1^2 + 4.3^2}}{4}$ .... (1 mark)

b) $\dfrac{1}{4} \times 5.2 \times 3.1 + \dfrac{(3.1^2 + 7.2^2)}{2}$ .... (1 mark)

c) $\dfrac{2}{5} \times \dfrac{\sqrt{3.1^2 - 1.1^2}}{4} + 12^2$ .... (1 mark)

Score   / 11

Total score   / 33

How well did you do? ✗ 1–10 **Try again**   11–19 **Getting there**   20–26 **Good work**   27–33 **Excellent!** ✓

For more help on this topic see KS3 Maths 5–8 Success Guide pages 18–19.

21

Checking calculations

A

Answer all parts of all questions.

1 Round each of these values to 3 significant figures.

a) 0.073 156 ....

b) 276 479 ....

c) 5215 ....

d) 0.370 794 .... (4 marks)

2 **25 624 people attend a football match.**

Write this number to 2 significant figures.. (1 mark)

3 **For each of these calculations round the numbers to 1 significant figure and work out an approximate answer.**

a) 279 × 31... (1 mark)

b) 627 × 217... (1 mark)

c) 36.8 × 42.8... (1 mark)

d) 475 ÷ 22... (1 mark)

e) 630 ÷ 9.9.. (1 mark)

4 **Round each of the numbers to 1 significant figure and work out an approximate answer.**

a) $\dfrac{19+41}{3.1}$... (1 mark)

b) $\sqrt{3.1 \times 20.2}$.. (1 mark)

c) $\left(\dfrac{27.5}{2.9}\right)^2$ .... (1 mark)

5 **Write down suitable calculations you could use to check the answers to the following questions.**

a) 2472 ÷ 6 = 412... (1 mark)

b) 932 × 5 = 4660... (1 mark)

c) 460 + 52 = 512.. (1 mark)

d) 1054 − 307 = 747.. (1 mark)

6 **A train goes 502 km in 1.3 hours.**

By rounding to 1 significant figure, give a rough estimate of its speed in kilometres per hour.

.... (1 mark)

7 **Paint is sold in 8-litre tins.**

Gill needs 42 litres of paint. How many tins must she buy?

.... (1 mark)

Score **/ 19**

Ⓒ *Indicates that a calculator may be used*

Answer all parts of the questions.

1 The table below shows some numbers. Round each of the numbers to the given number of significant figures.

Number	3 significant figures	2 significant figures	1 significant figure
2.7364			
4275			
0.038 65			

(9 marks)

2 An athletics club is taking part in the Commonwealth Games in Manchester. 715 athletes are competing in the Games. They are going to travel by coach. Each coach can carry 49 people.

a) How many coaches do they need for the journey?

Show your working.. (1 mark)

b) Each coach costs £365 to hire.

What is the total cost of the coaches?   £... (1 mark)

c) How much does each person pay to share the cost of the coaches equally?

£ .... (1 mark)

3 a) Circle the best estimate for the answer to 62.57 ÷ 11.94.

   3   4   5   6   7   8   (1 mark)

b) Circle the best estimate for the answer to 49.3 × 19.9.

   900   1000   1100   1200   1300   (1 mark)

c) Estimate the answer to $\dfrac{41.4 \times 20.6}{3.1 + 4.85}$

   Give your answer to 1 significant figure. .... (1 mark)

d) Estimate the answer to $\dfrac{(807 \div 40.13)^2}{4.01 + 6.1}$

   Give your answer to 1 significant figure. .... (1 mark)

Score   / 16

Total score   / 35

How well did you do?   1–8 **Try again**   9–16 **Getting there**   17–26 **Good work**   27–35 **Excellent!** ✔

For more help on this topic see KS3 Maths 5–8 Success Guide pages 20–21.

Ratio

A

Answer all parts of all questions.

1 Read each of the following statements, then write down if they are true or false.

a) The ratio 20 : 10 is the same as the ratio 2 : 1. .... (1 mark)

b) The ratio 25 : 35 is the same as the ratio 4 : 5. .... (1 mark)

c) The ratio 21 : 27 is the same as the ratio 7 : 9. .... (1 mark)

2 Write down each of the following ratios in the form 1 : *n*.

a) 20 : 30 ....   b) 3 : 5 ....   c) 6 : 2 ....   (3 marks)

3 Divide each amount in the given ratio.

a) £20 in the ratio 2 : 3 .... (1 mark)

b) £600 in the ratio 5 : 7 .... (1 mark)

4 In a wood there are 300 trees. The ratio of oak to ash trees is 2 : 13.

How many ash trees are there in the wood? .... (1 mark)

5 A photograph of length 27 cm is to be enlarged in the ratio 3 : 7.

What is the length of the enlarged photograph? .... (1 mark)

6 A recipe needs the following ingredients for 4 people.

Complete the recipe for 6 people.

4 people
150 g flour
2 eggs
200 g sugar

6 people
.... g flour
.... eggs
.... g sugar

(2 marks)

7 Seven bottles of cola cost £10.15. Work out the cost of twelve bottles of cola.

Ⓒ .... (2 marks)

8 Toothpaste is sold in three different sized tubes. Ⓒ

50 ml = £1.24   75 ml = £1.96   100 ml = £2.42

Which of the tubes of toothpaste gives the best value for money?

You must show full working out in order to justify your answer.

.... (3 marks)

Score   / 17

Ⓒ *Indicates that a calculator may be used*

Answer all parts of the questions.

1 a) One morning, Ahmed carried out a survey of the cars in a car park. He saw 10 red, 20 blue, 35 black and 50 silver cars. Complete the line below to show the ratios:

10 : 20 : .... : ....

2 : .... : .... : ....

(2 marks)

b) On another morning he carried out the survey again. He surveyed the same number of cars. The ratio of the cars this time is:

1 : 5 : 12 : 5

Complete the table, which shows the number of cars of each colour that are in the car park.

Colour	Number
Red	
Blue	
Black	
Silver	

(4 marks)

2 A tin holds 200 g of baked beans.
© The label on the tin shows this information:

Nutritional Information		200 g tin	
Energy	150 kcal	Carbohydrate	27.3 g
Protein	9.4 g	Fat	0.4 g

a) How many grams of carbohydrate would a 450 g tin of baked beans provide?

Show your working.. (2 marks)

b) On a different brand of baked beans, different information is shown:

Nutritional Information		250 g tin	
Energy	165 kcal	Carbohydrate	29.0 g
Protein	10.8 g	Fat	0.62 g

A girl eats the same amount of baked beans from both tins.
Which tin provides her with more protein?

Show your working.. (2 marks)

Score   / 10

Total score   / 27

How well did you do?   ✗ 1–6 Try again   7–11 Getting there   12–17 Good work   18–27 Excellent! ✓

For more help on this topic see KS3 Maths 5–8 Success Guide pages 22–23.

25

Indices

A

Choose just one answer, a, b, c or d.

1 In index form, what is the value of $5^3 \times 5^8$?   (1 mark)

a) 5^{24} ☐   b) 5^5 ☐   c) 2^5 ☐   d) 5^{11} ☐

2 In index form, what is the value of $(3^5)^2$?   (1 mark)

a) 10^3 ☐   b) 3^{10} ☐   c) 3^7 ☐   d) 3^3 ☐

3 What is the value of 4^0?   (1 mark)

a) 1 ☐   b) 4 ☐   c) 0 ☐   d) 16 ☐

4 What is the value of 3^{-2}?   (1 mark)

a) $\frac{1}{9}$ ☐   b) -9 ☐

c) 9 ☐   d) -6 ☐

5 What is the value of $7^{10} \div 7^{-2}$ written in index form?   (1 mark)

a) 7^8 ☐   b) 7^{12} ☐

c) 7^{20} ☐   d) 7^{14} ☐

Score   / 5

B

Answer all parts of all questions.

1 For each of the following, decide whether the expression is true or false.

a) $a \times a \times a = a^3$ ....

b) $6 \times 6 \times 6 \times 6 = 6^4$ ....

c) $y \times y^3 \times y^2 = y^7$ ....

d) $c^2 \times 3 \times c^4 \times 4 = 12c^6$ ....

e) $3 \times x \times x^2 \times 5 \times y^3 = 8x^3y^3$ ....   (5 marks)

2 LEVEL 8 Here are some cards with some expressions. Match each of the cards to the correct expression.

$5n^2$ $3n$ $6n^4$ $9n^5$ $4n^{-2}$

a) $3n \times 2n^3$ = ...

c) $12n^2 \div 4n$ = ...

e) $16n^4 \div 4n^6$ = ...

b) $27n^9 \div 3n^4$ = ...

d) $5 \times n^2$ = ...   (5 marks)

3 LEVEL 8 Simplify the following expressions.

a) $2x \times 3x$ = ...

c) $7m^3 \times 3m^4$ = ...

e) $16x^9 \div 4x^7$ = ...

g) $\frac{12a^2b}{4a}$ = ...

b) $9x^2 \times 4x^3$ = ...

d) $12x^4 \div 3x$ = ...

f) $16x \div 4x^3$ = ...

h) $\frac{25a^3b^2}{5ab}$ = ...   (8 marks)

4 LEVEL 8 Write these fractions using negative indices.   (4 marks)

a) $\frac{1}{10^3}$ b) $\frac{2}{x^3}$ c) $\frac{3}{a^5}$ d) $\frac{1}{2}$ Score   / 22

Answer all parts of the questions.

1 LEVEL 8 **Look at the table:**

4^0	=	1
4^1	=	4
4^2	=	16
4^3	=	64
4^4	=	256
4^5	=	1024
4^6	=	4096
4^7	=	16 384
4^8	=	65 536

a) Explain how the table shows that $64 \times 256 = 16\,384$. .... (1 mark)

b) Use the table to help you work out the value of $\frac{16\,384}{16}$. .... (1 mark)

c) Use the table to help you work out the value of $\frac{65\,536}{64}$. .... (1 mark)

d) The units digit of 4^5 is 4. What is the units digit of 4^{10}? .... (1 mark)

2 LEVEL 8

Some expressions are written on cards:

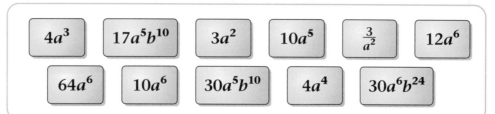

$4a^3$ $17a^5b^{10}$ $3a^2$ $10a^5$ $\frac{3}{a^2}$ $12a^6$

$64a^6$ $10a^6$ $30a^5b^{10}$ $4a^4$ $30a^6b^{24}$

Choose the correct card to go with each of the expressions below when fully simplified.

a) $5a^2 \times 2a^3$ .... b) $12a^4 \div 3a$ ....

c) $9a^3 \div 3a^5$ .... d) $(4a^2)^3$ ....

e) $15a^2b^4 \times 2a^3b^6$ .... (5 marks)

3 LEVEL 8 a) Simplify the expression $\frac{a^4b^3}{a^2b^2}$. .... (1 mark)

b) Simplify the expression $\frac{a^5b^2 \times a^3b^2}{a^2b^2}$. .... (1 mark)

c) Show that $\frac{(a+b)}{(a-b)(a+b)^2} + \frac{(a-b)}{(a-b)^2(a+b)}$

simplifies to $\frac{2}{(a+b)(a-b)}$. .... (2 marks)

Score   / 13

Total score   / 40

How well did you do?   ✗ 1–10 **Try again**   11–20 **Getting there**   21–30 **Good work**   31–40 **Excellent!** ✓

For more help on this topic see KS3 Maths 5–8 Success Guide pages 24–25.

Standard index form

A

Answer all parts of all questions.

1 **LEVEL 8** Write these numbers in standard form:

a) 2 670 000 = ...

b) 0.03296 = ...

c) 0.000 059 5 = ...

d) 2 million = ... (4 marks)

2 **LEVEL 8** Write as ordinary numbers:

a) 2.5×10^4 = ...

b) 6.02×10^{-3} = ...

c) 9.03×10^6 = ...

d) 1.27×10^{-5} = ... (4 marks)

3 **LEVEL 8** For each of the questions below, decide whether the statements are true or false.

a) 2700 written in standard form is 2.7×10^3. ....

b) 0.036 written in standard form is 3.6×10^2. ....

c) 0.000 27 written in standard form is 27×10^{-5}. ....

d) 251 000 written in standard form is 2.51×10^5. .... (4 marks)

4 **LEVEL 8** Carry out the following calculations, giving your answers in standard form.

a) $(3 \times 10^6) \times (2 \times 10^4)$ = ...

b) $(5 \times 10^7) \times (3 \times 10^2)$ = ...

c) $(9 \times 10^5) \div (3 \times 10^2)$ = ...

d) $(1.6 \times 10^9) \div (4 \times 10^{-2})$ = ... (4 marks)

5 **LEVEL 8** Work these out, giving your answers in standard form.

Ⓒ a) $(4.2 \times 10^8) \times (6.1 \times 10^5)$ = ...

b) $(9.3 \times 10^5) \times (3.6 \times 10^4)$ = ...

c) $(3.6 \times 10^{-4}) \times (3.8 \times 10^{-9})$ = ...

d) $(8.1 \times 10^{-2}) \times (4.2 \times 10^{-3})$ = ...

e) $(6.4 \times 10^9) \div (3.1 \times 10^4)$ = ...

f) $(6.9 \times 10^5) \div (2 \times 10^{-6})$ = ... (6 marks)

6 **LEVEL 8** The mass of a hair is 0.000 042 grams.

Ⓒ a) Write this number in standard form. .... (1 mark)

b) Calculate, in standard form, the mass of 6×10^5 hairs. .... (1 mark)

Score / 24

Ⓒ *Indicates that a calculator may be used*

B

Answer all parts of the questions.

1 **LEVEL 8** Charlotte is carrying out a science project.
C She has found out the diameters of two planets in centimetres.

Planet A 1.3×10^9 cm
Planet B 3.5×10^8 cm

a) How many times bigger is the diameter of planet A than the diameter of planet B?

Show your working.. (1 mark)

b) What is the difference in diameter between planet A and planet B?........................... (1 mark)

c) The formula $4\pi r^2$, where r is the radius of the planet, is used to work out the surface area.

   i) Work out the radius of planet A. ....cm (1 mark)

   ii) Work out the surface area of planet A, giving your answer in standard form.

      Show your working. ....cm^2 (2 marks)

2 **LEVEL 8** $(3 \times 10^3) \times (2 \times 10^4)$ can be written simply as 6×10^7.
Write these values as simply as possible:

a) $(5 \times 10^3) \times (1 \times 10^6) =$ .... (1 mark)

b) $\dfrac{8 \times 10^9}{4 \times 10^{-2}} =$ .... (1 mark)

3 **LEVEL 8** The mass of an atom is 2×10^{-23} grams.
C What is the total mass of 9×10^{15} of these atoms?

Show your working. ....grams (2 marks)

4 **LEVEL 8** A building is sold for £6.2 million.
C a) Write down 6.2 million in standard form. .... (1 mark)

b) Seven investors buy the building jointly. How much does each investor pay?
   Give your answer to 1 significant figure and write it in standard form. .... (2 marks)

Score   / 12

Total score   / 36

How well did you do?   ✗ 1–9 Try again   10–18 Getting there   19–26 Good work   27–36 Excellent! ✓

For more help on this topic see KS3 Maths 5–8 Success Guide pages 26–27.

29

STANDARD INDEX FORM   Number

Algebra 1

A   Answer all parts of all questions.

1   Write the following in symbols.

a) 4 more than t ....   b) 6 less than a ....

c) 5 times a ....   d) a divided by c ....

e) 5 less than 4 times b ....   f) 6 more than a, divided by n ....

g) 4 less than half of y ....   h) c less than a quarter of p ....

(8 marks)

2   Simplify the following expressions.

a) $a \times a \times a$ ....   b) $2 \times a \times 3 \times b$ ....

c) $a \times 3 \times a \times 4$ ....   d) $6 \times b^2 \times c \times b \times 2$ ....   (4 marks)

3   Here are some cards with expressions.

Match each expression with the equivalent statement.

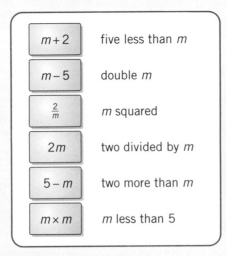

$m+2$	five less than m
$m-5$	double m
$\frac{2}{m}$	m squared
$2m$	two divided by m
$5-m$	two more than m
$m \times m$	m less than 5

(6 marks)

4   Decide whether these expressions, which have been simplified, are true or false.

a) $5a + 2a = 7a$ ....   b) $6a + 2b - b = 6a - b$ ....

c) $3a - 5b + 2b = 3a + 3b$ ....   d) $2a + 3a - 5a + 6b - b = 5b$ ....

(4 marks)

5   For an algebra pyramid, the next layer is formed by adding the two boxes in the row below.

Complete this algebra pyramid.
Write each expression as simply as possible.

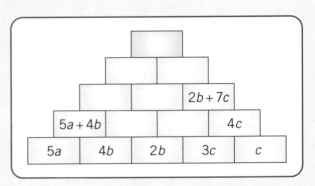

(3 marks)

Score   / 25

Answer all parts of the questions.

1 Write an expression for the perimeter of each of these shapes.
Write each expression in its simplest form.

a)

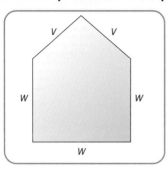

$p =$ .... (1 mark)

b)

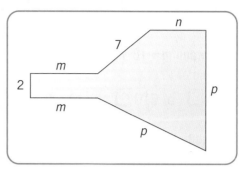

$p =$ .... (1 mark)

2 These patterns are made with matchsticks:

pattern 1 needs
6 matches

pattern 2 needs
11 matches

pattern 3 needs
16 matches

pattern 4 needs
21 matches

A rule to find how many matches are needed for each pattern is: $m = 5p + 1$

m stands for the number of matches.

p stands for the pattern number.

a) Use the rule to find the number of matches that would be needed in pattern number 10.

Show your working. .... (2 marks)

b) If 151 matches are used for another pattern, what would its pattern number be?

Show your working. .... (2 marks)

A different pattern is then made with the matches.

pattern 1 needs
5 matches

pattern 2 needs
9 matches

pattern 3 needs
13 matches

If p stands for the pattern number and m stands for the number of matches,
write down a rule that connects p and m. .... (2 marks)

Score   / 8

Total score   / 33

How well did you do?   ✗ 1–10 Try again   11–19 Getting there   20–26 Good work   27–33 Excellent! ✓

For more help on this topic see KS3 Maths 5–8 Success Guide pages 30–31.

Algebra 2

A

Choose just one answer, a, b, c or d.

1 If $v = mr$ and $m = 10$ and $r = 6$, what is the value of v? (1 mark)

a) 106 ☐ b) 610 ☐ c) 60 ☐ d) 16 ☐

2 When multiplied out and simplified, the expression $3(5x - 4)$ becomes: (1 mark)

a) $15x$ ☐   b) $15x - 12$ ☐

c) $5x - 12$ ☐   d) $15x - 4$ ☐

3 Factorise fully the expression $16x + 8$. (1 mark)

a) $2(8x + 4)$ ☐   b) $8(2x + 1)$ ☐

c) $8(2x)$ ☐   d) $4(4x + 2)$ ☐

4 Multiplied out and fully simplified, $(y - 2)^2$ is: (1 mark)

a) $y^2 - 2$ ☐   b) $y^2 + 4$ ☐

c) $y^2 - 4y + 4$ ☐   d) $y^2 + 4y + 4$ ☐

5 L8 Factorise the expression $n^2 + 4n + 3$. (1 mark)

a) $(n - 1)(n - 3)$ ☐ b) $n(n + 4 + 3)$ ☐

c) $(n - 2)(n + 1)$ ☐ d) $(n + 1)(n + 3)$ ☐

6 L8 Factorise the expression $y^2 - 25$. (1 mark)

a) $(y + 5)(y - 10)$ ☐ b) $(y - 5)(y - 5)$ ☐

c) $(y - 5)(y + 5)$ ☐ d) $(y - 5)(y - 10)$ ☐

Score   / 6

B

Answer all parts of all questions.

1 If $V = IR$:

a) Calculate V when $I = 24$ and $R = 2.5$ .... (1 mark)

b) Calculate V when $I = 2.7$ and $R = 10$ .... (1 mark)

c) Calculate R when $V = 65$ and $I = 5.2$ .... (1 mark)

2 Expand and simplify the following expressions.

a) $2x(x + 1)$ .... (1 mark)

b) $(n + 2)(n + 3)$ .... (1 mark)

c) $(n - 3)(n - 4)$ .... (1 mark)

d) $(n - 3)^2$ .... (1 mark)

3 Factorise the following expressions.

a) $10n + 20$ ....   b) $12 - 24n$ ....

c) $15n + 30$ ....   (3 marks)

4 Rearrange each of the formulae below to make n the subject.

a) $a = bn$ .... (1 mark)

b) $r = 2n - 4$ .... (1 mark)

c) $a = \frac{3n}{2} + 1$ .... (1 mark)

d) $y = 2(n + 1)$ .... (1 mark)

Score   / 14

Ⓒ *Indicates that a calculator may be used*

C

Answer all parts of the questions.

1 a) Nilam uses this formula to calculate the perimeter (P) of a shape.

$$P = a + 2b + \frac{\sqrt{a^2 + b^2}}{4}$$

Work out the perimeter of the shape if $a = 3$ and $b = 4.2$.

Show your working. .... (2 marks)

b) Naeve has a different shape. The formula to calculate the area (A) of the shape is given by:

$$A = r^2 + s^2 + \frac{4rs}{5}$$

Calculate the area of Naeve's shape when $r = 6$ and $s = 2.6$.

Show your working. .... (2 marks)

2 a) Two of the expressions below are equivalent.
Circle them.

$3(y + 4)$ $\qquad\qquad$ $2(4y + 3)$ $\qquad\qquad$ $4(2y + 4)$

$3(y + 2)$ $\qquad\qquad$ $2(4y + 8)$ $\qquad\qquad$ $2(4y + 5)$ $\qquad\qquad$ (1 mark)

b) Factorise this expression.

$6y + 12$ .... (1 mark)

c) Factorise fully this expression.

$4y^3 - 8y^2$ .... (2 marks)

d) Rearrange this formula to make y the subject.

$n = 4y - 3$ .... (1 mark)

e) **LEVEL 8** Fully factorise the following expressions.

i) $p^2 + 3p - 10$ .... $\qquad$ ii) $p^2 - 6p + 9$ ....

iii) $n^2 - 64$ .... $\qquad$ iv) $n^2 + 6n + 5$ .... (4 marks)

Score $\qquad$ / 13

Total score $\qquad$ / 33

How well did you do? ✗ **1–8 Try again** $\quad$ **9–15 Getting there** $\quad$ **16–25 Good work** $\quad$ **26–33 Excellent!** ✓

For more help on this topic see KS3 Maths 5–8 Success Guide pages 32–33.

Equations 1

A

Choose just one answer, a, b, c or d.

1 Solve the equation $n - 3 = 7$. (1 mark)
   a) 12 ☐   b) 2 ☐   c) 4 ☐   d) 10 ☐

2 Solve the equation $3n - 1 = 11$. (1 mark)
   a) 12 ☐   b) 36 ☐   c) 6 ☐   d) 4 ☐

3 Solve the equation $5n + 4 = 19$. (1 mark)
   a) 20 ☐   b) 3 ☐   c) 10 ☐   d) 5 ☐

4 Solve the equation $3(x + 2) = 12$. (1 mark)
   a) 8 ☐   b) 6 ☐   c) 2 ☐   d) 4 ☐

5 Solve the equation $2(3 - 2y) = 3y - 8$. (1 mark)
   a) 10 ☐   b) 5 ☐   c) 2 ☐   d) 6 ☐

Score   / 5

B

Answer all parts of all questions.

1 Solve the following equations.

   a) $3n + 2 = 14$ ....
   b) $\frac{n}{5} + 1 = 6$ ....
   c) $5n + 3 = 18$ ....
   d) $7n - 1 = 13$ ....
   e) $\frac{n}{3} - 4 = 2$ ....

   (5 marks)

2 Solve the following equations.

   a) $5n + 1 = 3n + 7$ ....
   b) $12n - 4 = 8n + 12$ ....
   c) $10n + 3 = 7n + 5$ ....
   d) $15n - 1 = 3n + 11$ ....

   (4 marks)

3 Solve the following equations.

   a) $5(3x - 1) = 20$ ....
   b) $7(2x + 3) = 28$ ....
   c) $4(n + 5) = 3(2n + 10)$ ....
   d) $2(n - 5) = 4(n + 2)$ ....

   (4 marks)

4 Solve the following equations.

   a) $5n + 6 = 3(n + 1)$ ....
   b) $2n - 3 + n = 2(n + 1)$ ....

   (2 marks)

5 The angles in a triangle add up to 180°.
Form an equation in n and solve it for each of the shapes. (6 marks)

a)
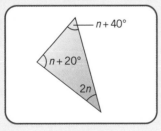

b)
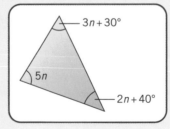

c)
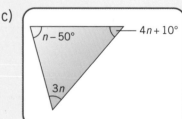

$n = $ ....
$n = $ ....
$n = $ ....

Score   / 21

C

Answer all parts of the questions.

1 In this wall, each brick is made by adding the values of the bricks beneath it.

a) Fill in the missing expressions on this wall.

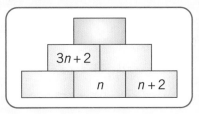

(3 marks)

b) If the value of the top brick is 24, work out the value of *n*.

n = ... (1 mark)

2 This L-shape is made with 8 square tiles:

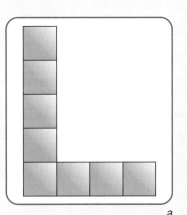

This is a square tile. The edge of the tile is *a* centimetres long: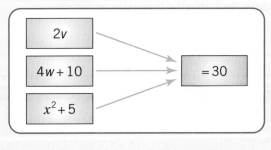

a) The perimeter of the L shape is 36 centimetres.
Write an equation involving *a*.

.... (1 mark)

b) Solve your equation to find the value of *a*.

a = ... (1 mark)

Solve these equations to find the values of *v*, *w* and *x*.

$2v$

$4w+10$ = 30

x^2+5

v = ...   w = ...   x = ...   (3 marks)

Score   / 9

Total score   / 35

For more help on this topic see KS3 Maths 5–8 Success Guide pages 34–35.

Equations 2 & inequalities

A

Answer all parts of all questions.

1 Solve the following simultaneous equations.

a) $x + y = 5$

$2x - y = 4$ ....

b) $3x + 2y = 7$

$5x - y = 3$ ....

c) $2x + y = 8$

$3x - y = 2$ ....

d) $5x + 3y = 34$

$4x - 3y = 11$ ....

e) $7x + 4y = 26$

$3x + 3y = 15$ ....

f) $3x + 4y = 19$

$4x - 3y = -8$ .... (12 marks)

2 a) On the axes provided,
draw the lines with equations:

   i) $x + y = 4$ Label this line P.

   ii) $y = x - 2$ Label this line Q.

b) Use the graphs to solve
the simultaneous equations:

$x + y = 4, \qquad y = x - 2$

$x =$ ....

$y =$ ....

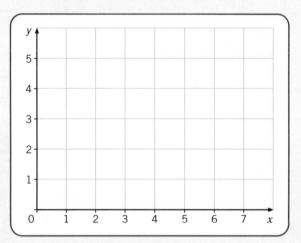

(2 marks)

(2 marks)

3 Using a trial and improvement method, solve the following equations,
C giving your answer to 1 decimal place.

a) $a^2 - 104 = 0$ .... (1 mark)

b) $a^3 - a = 15$ .... (1 mark)

4 On the number lines provided, draw the following inequalities. (2 marks)

a) $x > -3$

b) $2 < x \leqslant 5$ ....

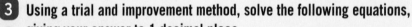

Solve the following inequalities.

5 a) $5x + 2 < 12$ ....

b) $\frac{x}{3} + 1 \geqslant 3$ ....

c) $3 \leqslant 2x + 1 \leqslant 9$ ....

d) $3 \leqslant 3x + 2 \leqslant 8$ .... (4 marks)

Score   / 24

© *Indicates that a calculator may be used*

B

Answer all parts of the questions.

1 Annabelle makes a rectangle with an area of 24.5 cm².
The sides of Annabelle's rectangle are
n cm and $(10 - n)$ cm.

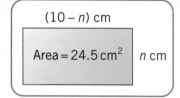

$(10 - n)$ cm

Area = 24.5 cm² n cm

She wants to find a value of n, to one decimal place, which gives her an area as close as possible to 24·5 cm².

Use the table to find the value of n, to one decimal place.

n	$10 - n$	Area
2	8	16

$n = $ .... (1 d.p.)

(3 marks)

2 Solve these simultaneous equations to find the values of a and b.

Show your working.

$a = $ ....

$b = $ ....

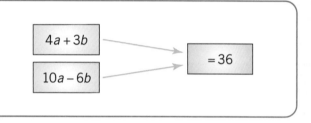

$4a + 3b$

$= 36$

$10a - 6b$

(4 marks)

3 **LEVEL 8** a) On the grid below, draw the graph $x + y = 4$.

Shade in the region where $x + y \leqslant 4$. (2 marks)

LEVEL 8 b) On the same grid, draw the graph $y - x = 1$.

Shade in the region where $y - x \geqslant 1$. (2 marks)

LEVEL 8 c) Label the region R where both these inequalities are satisfied. (1 mark)

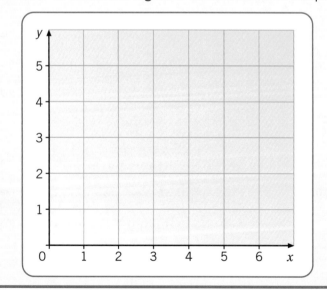

Score   / 12

Total score   / 36

How well did you do?   ✗ 1–9 Try again   10–16 Getting there   17–27 Good work   28–36 Excellent! ✓

For more help on this topic see KS3 Maths 5–8 Success Guide pages 36–39.

Patterns & sequences

A

Choose just one answer, a, b, c or d.

1 The rule for a sequence is: 'multiply the previous number by 2 and subtract 3'. If the first term in the sequence is –2, what is the next term? **(1 mark)**

a) –1 ☐   b) 1 ☐   c) –7 ☐   d) 7 ☐

2 What is the next number in this sequence? **(1 mark)**

3, 7, 11, 15

a) 20 ☐   b) 17 ☐   c) 18 ☐   d) 19 ☐

3 What is the nth term of the sequence of which the first four terms are 3, 7, 11, 15? **(1 mark)**

a) $n + 4$ ☐   b) $4n + 1$ ☐

c) $3n + 2$ ☐   d) $4n - 1$ ☐

4 Here are the first four terms of a sequence:

2, 5, 10, 17

What is the nth term of this sequence? **(1 mark)**

a) $n + 3$ ☐   b) $n^2 + 1$ ☐

c) n^2 ☐   d) $n + 5$ ☐

5 If the nth term of a sequence is given by $5n - 2$, what is the third term of the sequence? **(1 mark)**

a) 13 ☐

b) 15 ☐

c) 3 ☐

d) 17 ☐

Score   / 5

B

Answer all parts of all questions.

1 Grace has written a sequence of numbers: 4, 7, 10, 13

a) Write down the next three terms in Grace's sequence.. **(1 mark)**

b) What is the 10th term of this sequence?.. **(1 mark)**

c) What is the nth term of Grace's sequence?... **(1 mark)**

2 Look at this sequence: 5, 9, 13, 17

a) What is the 6th number in this sequence? .... **(1 mark)**

b) Describe how this sequence is formed. .... **(1 mark)**

c) Write down the nth term of this sequence. .... **(1 mark)**

3 Look at this sequence of numbers: 3, 8, 15, 24

a) What is the next number in the sequence?... **(1 mark)**

b) Write down the nth term of this sequence. .... **(1 mark)**

c) What is the 12th term of this sequence?.. **(1 mark)**

4 For each of these sequences, decide whether the nth term given is correct.

a) 2, 4, 6, 8, 10 nth term: $n + 2$ .... **(1 mark)**

b) 1, 4, 7, 10, 13 nth term: $3n - 2$ .... **(1 mark)**

c) 1, 4, 9, 16, 25 nth term: n^2 .... **(1 mark)**

Score   / 12

C

Answer all parts of the questions.

1 a) This is a number sequence: 5, 8, 11, 14, ...

Write down the *n*th term of this number sequence.

.... (1 mark)

b) This is another number sequence: 1, 4, 9, 16, ...

Write down the *n*th term of this number sequence.

.... (1 mark)

2 **Numbered squares are piled up. The diagram shows the first 4 rows.**

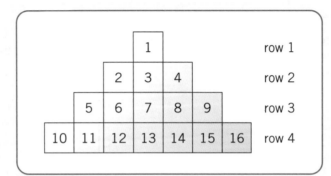

a) The last number in row *n* is n^2.

For example, the last number in row 2 is $2^2 = 4$.

i) Calculate the last number in row 7. .... (1 mark)

ii) Calculate the first number in row 8. .... (1 mark)

b) The last number in row *n* is n^2.

Write an expression for the number that is two numbers before the last one in row *n*.

.... (1 mark)

c) The first number in row *n* is $(n - 1)^2 + 1$.

Calculate the first number in the 10th row. .... (1 mark)

d) Write down an expression as simply as possible for the second number in row *n*.

.... (1 mark)

e) **LEVEL 8** Multiply out and simplify the expression $(n - 1)^2 + 1$. .... (1 mark)

Score   / 8

Total score   / 25

How well did you do?   ✗ **1–6 Try again   7–11 Getting there   12–17 Good work   18–25 Excellent!** ✓

For more help on this topic see KS3 Maths 5–8 Success Guide pages 40–41.

Coordinates & graphs

A

Answer all parts of all questions.

1 a) Complete the table of values for each of these lines.

i) $y = 2x$

x	−2	−1	0	1	2	3
y						

ii) $y = 3x - 2$

x	−1	0	1	2	3
y					

(2 marks)

b) Draw the graphs
$y = 2x$ and $y = 3x - 2$
on the grid opposite.

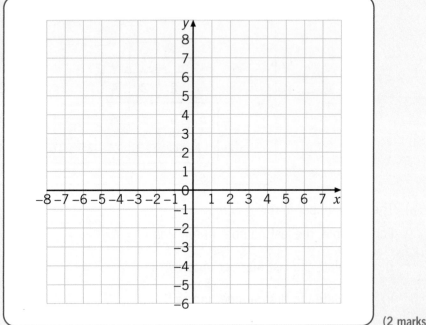

(2 marks)

2 The diagram below shows some straight-line graphs. Write down the equation of each line.

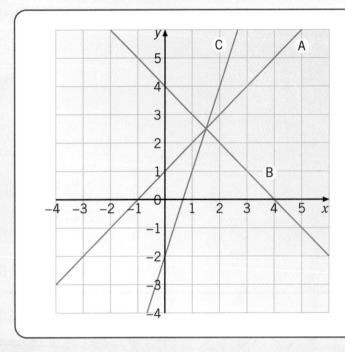

Line A: .... (1 mark)

Line B: .... (1 mark)

Line C: .... (1 mark)

Score   / 7

Answer all parts of the questions.

1 The diagram shows the graph of the line $y = 3x$.

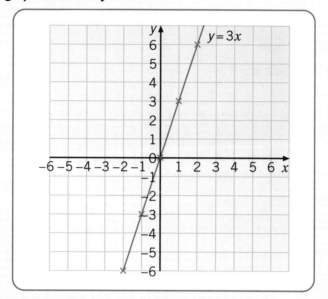

a) The straight line with the equation $y = 2x - 1$ goes through the point (3, 5).

i) Work out two more coordinates that lie on the line $y = 2x - 1$.

(................, ...)   (................, ...)   (2 marks)

ii) On the diagram above, draw the graph of the straight line $y = 2x - 1$.
Label your line $y = 2x - 1$.   (2 marks)

b) Write the equation of a straight line that goes through the point (0, –4) and is parallel to the straight line $y = 3x$.

$y =$ .... (1 mark)

2 Here are five different equations, labelled A to E.

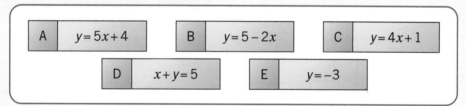

A $y = 5x + 4$ B $y = 5 - 2x$ C $y = 4x + 1$

D $x + y = 5$ E $y = -3$

Think about the graphs of these equations.

a) Which graph goes through the point (0, 1)? .... (1 mark)

b) Which graph is parallel to the x-axis? .... (1 mark)

c) Which graph is parallel to $y = 5x - 2$? .... (1 mark)

d) Which two graphs meet at the point (0, 5)? .... (2 marks)

Score   / 10

Total score   / 17

How well did you do?   ✗ 1–3 Try again   4–8 Getting there   9–12 Good work   13–17 Excellent!   ✓

For more help on this topic see KS3 Maths 5–8 Success Guide pages 42–43.

More graphs

A

Answer all parts of all questions.

1 Complete the tables for the graphs of:

a) i) $y = x^2$

x	–3	–2	–1	0	1	2	3
y			1			4	

ii) $y = 2x^2 – 4$

x	–3	–2	–1	0	1	2	3
y	14			–4		4	

b) Draw the graphs on the grid opposite.

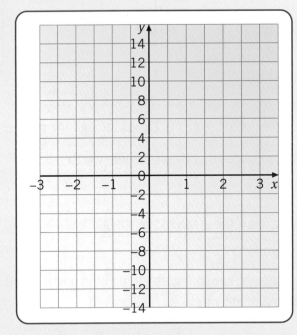

(4 marks)

2 Match each of the sketch graphs to the equations below.

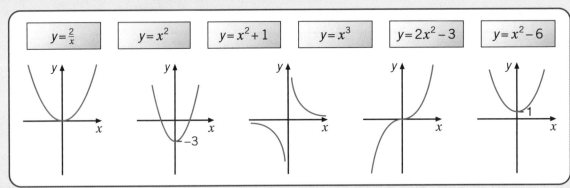

$y = \frac{2}{x}$ $y = x^2$ $y = x^2 + 1$ $y = x^3$ $y = 2x^2 – 3$ $y = x^2 – 6$

Equation:

a) ....   b) ....   c) ....   d) ....   e) ....

(5 marks)

3 For each of the functions below, state whether it is a linear or a quadratic function.

a) $y = x^2 – 4$ ....

b) $y = 3 – 2x$ ....

c) $y = 6x – 1$ ....

d) $y = 4x^2 – 2x + 1$ ....

(4 marks)

Score   / 13

B

Answer all parts of the questions.

1 The table below shows the value of x and y for the equation $y = x^2 - 4$.

a) Complete the table.

x	-2	-1	0	1	2	3
y				-3	0	5

(2 marks)

b) On the grid opposite, draw the graph. Make sure you label the graph.

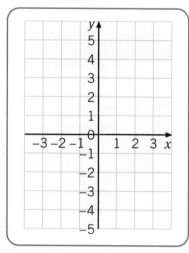

(2 marks)

c) On the same grid, sketch the graph of $y = x^2 + 2$. (1 mark)

d) Write down the equation of the line of symmetry of these graphs.

.... (1 mark)

2 Here are seven different equations labelled A to G:

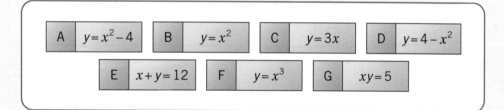

Think about the graphs of these equations.

a) Which three graphs go through the point (0, 0)?

...................................., ... and .... (3 marks)

b) Which graph goes through the point (3, 27)? .... (1 mark)

c) Which graphs are symmetrical about the y-axis? .... (1 mark)

d) Which graph goes through the point (1, 3)? .... (1 mark)

e) Which graph goes through the point (1, 5)? .... (1 mark)

Score   / 13

Total score   / 26

How well did you do?   ✗ 1–6 Try again   7–11 Getting there   12–17 Good work   18–26 Excellent! ✓

For more help on this topic see KS3 Maths 5–8 Success Guide pages 44–45.

Interpreting graphs

A Answer all parts of all questions.

1 This conversion graph converts British pounds (£) into euros (€) and vice versa.

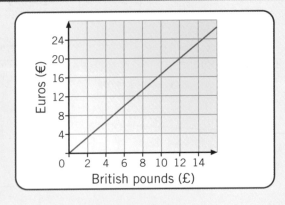

Use the graph to convert:

a) £10 into euros .... (1 mark)

b) €16 into pounds .... (1 mark)

2 Amy has a mobile phone. Her mobile has this tariff:

Monthly charge £10
Calls cost 10p per minute

a) Complete the table, which shows Amy's costs:

Number of minutes	0	10	20	30	40	50	60
Cost (£)	10			13			16

(2 marks)

b) Draw the graph showing Amy's costs on the axes provided.

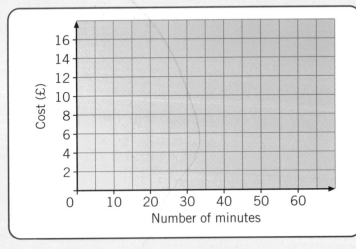

(1 mark)

c) How much would 35 minutes of calls cost?

.... (1 mark)

d) If Amy pays £25, for how many minutes did she use the phone that month?

.... (1 mark)

e) Explain why Amy's bill cannot be less than £10.

.... (1 mark)

Score /8

44

Letts

KS3
Success

**Workbook
Answer
Booklet**

Mathematics

Answers

NUMBERS

Pages 6–7 Numbers

A
1. b
2. d
3. c
4. d
5. a
6. a

B
1. a) 1, 4, 6, 12
   b) 13, 17
   c) 1, 4, 49
   d) 6, 12
2. 1 3 5 6 10 15
3. 24
4. 2
5. $2 \times 2 \times 3 \times 5$
6. False
7. 1, 2, 3, 4, 6, 8, 12, 24
8. a) $4^3 = 4 \times 4 \times 4$
   b) $2^5 = 2 \times 2 \times 2 \times 2 \times 2$

C
1. a) ○○○○○○
      ○○○○○○
   b) ○○○○○○○○○○○○
   c)

Number of rows	Number of counters in each row
1 row	12 counters in each row
2 rows	6 counters in each row
3 rows	4 counters in each row
4 rows	3 counters in each row
6 rows	2 counters in each row
12 rows	1 counter in each row

   d) 5 is not a factor of 12
2. Penelope's number is 16
   Tom's number is 2

Pages 8–9 Positive & negative numbers

A
1. a) –3 and –7
   b) 0 and –7
   c) 5 and –3
2. a) 12
   b) 8
   c) 12
   d) –3
   e) 3
   f) –4
3. a) 21
   b) –5
   c) –2
   d) –16
4. –22°C
5.

```
            4
         7    -3
      4    3    -6
   7   -3    6   -12
```

6. a) $-12 \div -3 = 4$
   b) $-7 - (-3) = -4$
   c) $9 \div -3 = -3$
   d) $12 + (-5) = 7$

B
1. a) Arrow points to 32°C
   b) Arrow points to –7°C
   c) –2°C
   d) 12°C
   e) 15°C, 9°C, 0°C, –3°C, –6°C, –12°C
2. a) –1
   b) –5
   c) 5
   d) –5
   e) 3
   f) –5

Pages 10–11 Working with numbers

A
1. a) A and C
   b) D
2.

6	1	8
7	5	3
2	9	4

3. a) 11 718
   b) 529 620
   c) 45
   d) 83
4. £26.64
5. 42
6. 42
7. £8190

B
1. a) 270 pencils
   b) £24.50
   c) £10.50
   d) 12 boxes
2. $170 - 110$
   $1200 \div 20$
   $\frac{1}{4}$ of 240
   $31 + 29$
3. a) £249.75
   b) (i) 33 shrubs
      (ii) 220 grams
4. a)

	4		0

   b)

4	2	7	0	0	0

Pages 12–13 Fractions

A
1. d
2. b
3. b
4. d

B
1. a) $\frac{31}{35}$
   b) $\frac{3}{26}$
2. a) $\frac{2}{9}, \frac{1}{3}, \frac{1}{2}, \frac{4}{7}, \frac{4}{5}$
   b) $\frac{2}{7}, \frac{1}{3}, \frac{3}{5}, \frac{9}{13}$
3. a) $\frac{8}{63}$
   b) 4
   c) $\frac{18}{55}$
   d) $2\frac{2}{3}$

4. a) True
   b) False
   c) False
   d) True
5. 1320
6. 70
7. 210

C
1. a) 4 cakes
   b) $\frac{12}{16} = \frac{3}{4}$
   c) 16 cakes
2. $\frac{4}{9}$ of 18
   $\frac{1}{5}$ of 40
   $\frac{2}{3}$ of 12
   One-half of 16
3. a) £30
   b) £600
   c) $\frac{31}{35}$
   d) $\frac{8}{35}$
4. a) $\frac{5}{16}$
   b) £26.25

Pages 14–15 Decimals

A
1. d
2. a
3. d
4. c
5. a

B
1. 2.71, 3.62, 3.69, 3.691, 4.38, 4.385, 4.39
2.

```
            34.5
        14.2    20.3
      6.3   7.9   12.4
   2.7   3.6   4.3   8.1
```

3. a) Rebecca
   b) 0.085
4. £69.48
5. a) True
   b) False
   c) False
6. a) 0.01   b) 0.001
   c) 10   d) 0.001
   e) 0.1   f) 0.001
   g) 0.01   h) 0.1

C
1. a) 6.4
   b) 16.6
   c) 281.25
   d) 15.6
2. a) $0.02 \times 0.01 = 0.0002$
   b) $3 \div 0.02 = 150$
   c) $10 \div 0.01 = 1000$

Pages 16–17 Percentages 1

A
1. £176 000
2. £110.50

3.

10% of 60		32
25% of 200		45
40% of 80		6
15% of 120		18
30% of 150		50

4. 93
5. £22.56
6. a) Maths   80%
      English   54%
      Geography   93%
      Science   81%
      German   38%
   b) German
7. a) 39%
   b) 80%

B
1. 10% of £80 = £8
   5% of £80 = £4
   $2\frac{1}{2}$% of £80 = £2
   $17\frac{1}{2}$% of £80 = £14
2. a) 41.7%
   b) 18.6%
   c) 12.9%
3. a) (i) £1.30
      (ii) £9.75
   b) £143.28

Pages 17–18 Percentages 2

A
1. c
2. d
3. d
4. b
5. d

B
1. 24%
2. £42.84
3. £10 288.50
4. £365
5. £4113.64
6. £89.47

C
1. a) 50.4%
   b) 25% increase
2. a) £435.20
   b) £812.50
   c) 5 days
3. a) 47%
   b) £5230.77

Pages 20–21 Equivalents & using a calculator

A
1. a) (i) 0.6   (ii) 60%
   b) (i) 0.01   (ii) 1%
   c) (i) 0.125   (ii) 12.5%
2. a) 0.3, $\frac{3}{10}$, 30%
   b) 0.25, $\frac{1}{4}$, 25%
   c) 0.1, $\frac{1}{10}$, 10%
   d) 0.45, $\frac{9}{20}$, 45%

3. 25%, 30%, $\frac{3}{8}$, $\frac{2}{5}$, 0.41, 72%, 0.9
4. Supers = £380
   Electricals = £408
   Ramones = £450
   Supers is cheapest.
5. a) 11
   b) 36
   c) 15
6. a) 0.7317
   b) 0.2097
   c) 73.0974
      (all to 4 d.p.)

B

1. $\frac{2}{5}$, 0.4, 40%
   $\frac{1}{2}$, 0.5, 50%
   $\frac{1}{8}$, 0.125, 12.5%
   $\frac{15}{20}$, 0.75, 75%

2. 90%, 0.9, $\frac{9}{10}$
   33.$\dot{3}$%, 0.3$\dot{3}$, $\frac{1}{3}$
   37.5%, 0.375, $\frac{3}{8}$
   5%, 0.05, $\frac{1}{20}$

3. a) 13.5 (1 d.p.)
   b) 34.8 (1 d.p.)
   c) 144.3 (1 d.p.)

Pages 22–23 Checking calculations

A

1. a) 0.0732
   b) 276 000
   c) 5220
   d) 0.371
2. 26 000
3. a) 9000
   b) 120 000
   c) 1600
   d) 25
   e) 60
4. a) 20
   b) 8
   c) 100
5. a) 412 × 6 = 2472
   b) 4660 ÷ 5 = 932
   c) 512 − 52 = 460
   d) 747 + 307 = 1054
6. 500 km/h
7. 6 tins of paint.

B

1.

Number	3 s.f.	2 s.f.	1 s.f.
2.7364	2.74	2.7	3
4275	4280	4300	4000
0.038 65	0.0387	0.039	0.04

2. a) 15 coaches
   b) £5475
   c) £7.66
3. a) 5
   b) 1000
   c) 100
   d) 40

Pages 24–25 Ratio

A

1. a) True
   b) False
   c) True
2. a) 1 : 1.5
   b) 1 : 1.$\dot{6}$
   c) 1 : 0.$\dot{3}$
3. a) 8 : 12
   b) 250 : 350
4. 260 ash trees
5. 63 cm
6. 225 g flour
   3 eggs
   300 g sugar
7. £17.40
8. The 100 ml tube

B

1. a) 10 : 20 : 35 : 50
      2 : 4 : 7 : 10

Colour	Number
Red	5
Blue	25
Black	60
Silver	25

2. a) 61.43 g (2 d.p.)
   b) 200 g tin provides 0.047 grams of protein per gram of beans. 250 g tin provides 0.0432 g of protein per gram of beans, so the 200 g tin provides more protein.

Pages 26–27 Indices

A

1. d
2. b
3. a
4. a
5. b

B

1. a) True
   b) True
   c) False
   d) True
   e) False
2. a) $6n^4$
   b) $9n^5$
   c) $3n$
   d) $5n^2$
   e) $4n^{-2}$
3. a) $6x^2$
   b) $36x^5$
   c) $21m^7$
   d) $4x^3$
   e) $4x^2$
   f) $4x^{-2}$
   g) $3ab$
   h) $5a^2b$
4. a) 10^{-3}
   b) $2x^{-3}$
   c) $3a^{-5}$
   d) 2^{-1}

C

1. a) $64 \times 256 = 4^3 \times 4^4 = $
      $= 4^7 = 16\,384$

b) $\dfrac{16\,384}{16} = \dfrac{4^7}{4^2} = 4^5 = 1024$
c) $\dfrac{65\,536}{64} = \dfrac{4^8}{4^3} = 4^5 = 1024$
d) $4^{10} = (4^5)^2$
   $4^2 = 16$
   units digit is 6
2. a) $10a^5$
   b) $4a^3$
   c) $\dfrac{3}{a^2}$
   d) $64a^6$
   e) $30a^5 b^{10}$
3. a) a^2b
   b) a^6b^2
   c)
   $$\dfrac{(a+b)}{(a-b)(a+b)^2} + \dfrac{(a-b)}{(a-b)^2(a+b)}$$
   $$= \dfrac{1}{(a-b)(a+b)} + \dfrac{1}{(a-b)(a+b)}$$
   $$= \dfrac{2}{(a+b)(a-b)}$$

Pages 28–29 Standard index form

A

1. a) 2.76×10^6
   b) 3.296×10^{-2}
   c) 5.95×10^{-5}
   d) 2×10^6
2. a) 25 000
   b) 0.006 02
   c) 9 030 000
   d) 0.000 012 7
3. a) True
   b) False
   c) False
   d) True
4. a) 6×10^{10}
   b) 1.5×10^{10}
   c) 3×10^3
   d) 4×10^{10}
5. a) 2.562×10^{14}
   b) 3.348×10^{10}
   c) 1.368×10^{-12}
   d) 3.402×10^{-4}
   e) 2.06×10^5
   f) 3.45×10^{11}
6. a) 4.2×10^{-5}
   b) 2.52×10^1 grams

B

1. a) 3.7 times bigger
   b) 9.5×10^8 cm
   c) (i) 6.5×10^8 cm
      (ii) 5.3×10^{18} cm^2
2. a) 5×10^9
   b) 2×10^{11}
3. 1.8×10^{-7} grams
4. a) $6\,200\,000 = 6.2 \times 10^6$
   b) 9×10^5

ALGEBRA
Pages 30–31 Algebra 1

A

1. a) $t + 4$
   b) $a − 6$
   c) $5a$
   d) $\dfrac{a}{c}$

e) $4b − 5$
f) $\dfrac{a+6}{n}$
g) $\dfrac{y}{2} − 4$
h) $\dfrac{p}{4} − c$
2. a) a^3
   b) $6ab$
   c) $12a^2$
   d) $12b^3c$
3.

$m + 2$	five less than m
$m − 5$	double m
$\frac{2}{m}$	m squared
$2m$	two divided by m
$5 − m$	two more than m
$m \times m$	m less than 5

4. a) True
   b) False
   c) False
   d) True
5.

	$5a+28b$ $+13c$	
	$5a+18b$ $+3c$	$10b$ $+10c$
$5a+10b$	$8b+3c$	$2b+7c$
$5a+4b$	$6b$	$2b+3c$ $4c$
$5a$	$4b$	$2b$ $3c$ c

B

1. a) $p = 2v + 3w$
   b) $p = 2p + 2m + n + 9$
2. a) $5 \times 10 + 1 = 51$ matches
   b) pattern number 30
   c) $m = 4p + 1$

Pages 32–33 Algebra 2

A

1. c
2. b
3. b
4. c
5. d
6. c

B

1. a) 60
   b) 27
   c) 12.5
2. a) $2x^2 + 2x$
   b) $n^2 + 5n + 6$
   c) $n^2 − 7n + 12$
   d) $n^2 − 6n + 9$
3. a) $10(n + 2)$
   b) $12(1 − 2n)$
   c) $15(n + 2)$
4. a) $n = \dfrac{a}{b}$
   b) $n = \dfrac{r+4}{2}$
   c) $n = \dfrac{2(a-1)}{3}$
   d) $n = \dfrac{y}{2} − 1$

C

1. a) $P = 3 + (2 \times 4.2)$
   $\quad + \sqrt{\dfrac{(3^2 + 4.2^2)}{4}}$
   $P = 12.69$
   b) $A = 6^2 + 2.6^2$
   $\quad + \dfrac{(4 \times 6 \times 2.6)}{5}$
   $A = 55.24$
2. a) $4(2y + 4)$ and
   $2(4y + 8)$
   b) $6(y + 2)$
   c) $4y^2(y - 2)$
   d) $y = \dfrac{n + 3}{4}$
   e) i) $(p - 2)(p + 5)$
   ii) $(p - 3)(p - 3)$
   iii) $(n - 8)(n + 8)$
   iv) $(n + 5)(n + 1)$

Pages 34–35 Equations 1

A

1. d
2. d
3. b
4. c
5. c

B

1. a) $n = 4$
   b) $n = 25$
   c) $n = 3$
   d) $n = 2$
   e) $n = 18$
2. a) $n = 3$
   b) $n = 4$
   c) $n = \dfrac{2}{3}$
   d) $n = 1$
3. a) $x = \dfrac{5}{3}$
   b) $x = \dfrac{1}{2}$
   c) $n = -5$
   d) $n = -9$
4. a) $n = -\dfrac{3}{2}$
   b) $n = 5$
5. a) $n = 30°$
   b) $n = 11°$
   c) $n = 27.5°$

C

1. a)

	5n + 4	

3n + 2	2n + 2

2n + 2	n	n + 2

   b) $n = 4$
2. a) $18a = 36$
   b) $a = 2$
3. $v = 15$
   $w = 5$
   $x = \pm5$

Pages 36–37 Equations 2 & inequalities

A

1. a) $x = 3, y = 2$
   b) $x = 1, y = 2$
   c) $x = 2, y = 4$

d) $x = 5, y = 3$
e) $x = 2, y = 3$
f) $x = 1, y = 4$
2. a)

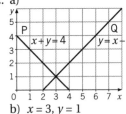

   b) $x = 3, y = 1$
3. a) $a = 10.2$
   b) $a = 2.6$
4. a)

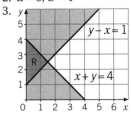

   b)

5. a) $x < 2$
   b) $x \geqslant 6$
   c) $1 \leqslant x \leqslant 4$
   d) $\dfrac{1}{3} \leqslant x \leqslant 2$

B

1. $n = 4.3\,\text{cm}$
2. $a = 6, b = 4$
3.

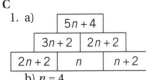

Pages 38–39 Patterns & sequences

A

1. c
2. d
3. d
4. b
5. a

B

1. a) 16, 19, 22
   b) 31
   c) $3n + 1$
2. a) 25
   b) Each term is the previous term plus 4.
   c) $4n + 1$
3. a) 35
   b) $n^2 + 2n$
   c) 168
4. a) False
   b) True
   c) True

C

1. a) $3n + 2$
   b) n^2
2. a) i) 49
   ii) 50
   b) $n^2 - 2$
   c) 82
   d) $(n - 1)^2 + 2 = n^2 - 2n + 3$
   e) $n^2 - 2n + 2$

Pages 42–43 Coordinates & graphs

A

1. a) i) $y = 2x$

x	-2	-1	0	1	2	3
y	-4	-2	0	2	4	6

   ii) $y = 3x - 2$

x	-1	0	1	2	3
y	-5	-2	1	4	7

   b)

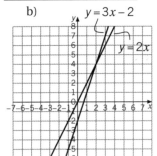

2. Line A: $y = x + 1$
   Line B: $y = -x + 4$
   Line C: $y = 3x - 2$

B

1. a) i) (2, 3) (0, -1)
   (or any two points on the line)

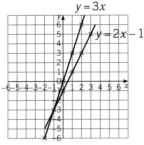

   ii) See graph.
   b) $y = 3x - 4$
2. a) C: $y = 4x + 1$
   b) E: $y = -3$
   c) A: $y = 5x + 4$
   d) B: $y = 5 - 2x$ and
   D: $x + y = 5$

Pages 42–43 More graphs

A

1. a) i) $y = x^2$

x	-3	-2	-1	0	1	2	3
y	9	4	1	0	1	4	9

   ii) $y = 2x^2 - 4$

x	-3	-2	-1	0	1	2	3
y	14	4	-2	-4	-2	4	14

   b)

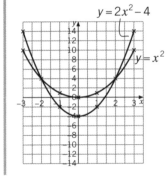

2. a) $y = x^2$
   b) $y = 2x^2 - 3$
   c) $y = \dfrac{2}{x}$
   d) $y = x^3$
   e) $y = x^2 + 1$
3. a) Quadratic function
   b) Linear function
   c) Linear function
   d) Quadratic function

B

1. a)

x	-2	-1	0	1	2	3
y	0	-3	-4	-3	0	5

   b) c)

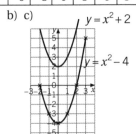

   d) $x = 0$
2. a) $y = x^2, y = 3x, y = x^3$
   b) $y = x^3$
   c) $y = x^2 - 4, y = x^2, y = 4 - x^2$
   d) $y = 4 - x^2$
   e) $xy = 5$

Pages 44–45 Interpreting graphs

A

1. a) approx. €16.50
   b) £9.60
2.

Number of minutes	0	10	20	30	40	50	60
Cost (£)	10	11	12	13	14	15	16

   b)

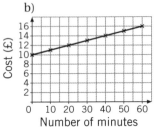

   c) £13.50
   d) 150 minutes
   e) Even if Amy does not use her phone, she has to pay the £10 standing charge. So her bill will never be less than £10.

B

1. Graph 1 matches container D
   Graph 2 matches container B
   Graph 3 matches container A

2. a) 30 mph
   b) Taxi has stopped.
   c)

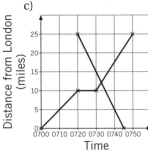

   d) 0733

SHAPE, SPACE AND MEASURES
Pages 46–47 Shapes
A
1. a is an isosceles triangle
   b is a scalene triangle
   c is an equilateral triangle
2. a) True
   b) True
   c) False
   d) True
   e) True
3. a) Parallel
   b) 2
   c) 4
   d) Circumference
   e) 90°
4.

   a) 6
   b) 6

B
1. a) Trapezium and quadrilateral
   b) Equilateral, 3, 60
2. a) Pentagon
   b) Quadrilateral
   c) Octagon
   d) Heptagon
3.

Pages 48–49 Solids
A
1. c
2. c
3. a
4. d
5. c

B
1. Nets a, b, e
   a) 12 edges
   b) 8 vertices
2. a)   b)

C
1. a)

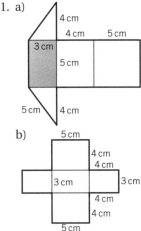

   b)

2. a) Side view 2
   Side view 1
   Side view 4
   Side view 3
   b)

Pages 50–51 Constructions & LOGO
A
1. Forward 6
   Turn right 120°
   Forward 6
2. Accurate diagrams to be drawn. Lengths accepted that are ± 2 mm.
3. Accurate diagrams to be drawn. Lengths to be accurate to ± 2 mm, angles to ± 2°.
4. Perpendicular bisector to be drawn at 3.5 cm.
5. Line bisects angle at 30°.

B
1. a) Forward 12
   Turn right 90°
   Forward 12
   Turn right 90°
   Forward 12
   Turn right 90°
   Forward 12
   b) Forward 5
   Turn right 60°
   Forward 3
   Turn right 120°
   Forward 5
   Turn right 60°
   Forward 3
2. Accurate diagrams to be drawn. Lengths accepted which are ±2 mm.

Pages 52–53 Loci & coordinates in 3D
A
1. a) True
   b) True
   c) False
   d) True

2.

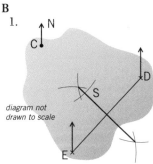

B
1.

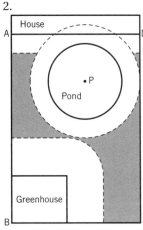

diagram not drawn to scale

2.

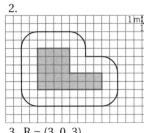

3. R = (3, 0, 3)
   S = (3, 3, 1)
   T = (0, 3, 1)

Pages 54–55 Angles & tessellations
A
1. a) 60°
   b) 80°
   c) 163°
   d) 138°
   e) 65°
   f) 102°
2. a) a = 130°
      b = 50°
      c = 50°
   b) a = 60°
      b = 60°
      c = 120°
3. 20 sides
4. 720°
5.

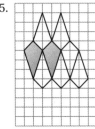

B
1. a) i) a = 80°
      ii) b = 20°
   b) a = 65°
      b = 95°
      c = 150°
   c) i) a = 60°
      b = 120°
   ii) Three hexagons tessellate at a point, since 3 × 120° = 360°. Shapes will only tessellate when the angles add up to 360°.

Pages 56–57 Bearings & scale drawings
A
1. a) (i) 072°   (ii) 252°
   b) (i) 310°   (ii) 130°
   c) (i) 230°   (ii) 050°
   d) (i) 150°   (ii) 330°
2. Accurate diagrams to be drawn. Lengths to be accurate to ±2 mm, angles to ±2°.
   Distance 9.8 km
   Bearing 240°
3. 10 km

B
1. a) Accurate diagrams to be drawn. Lengths to be accurate to ±2 mm, angles to ±2°.
   b) 9.5 cm
2. a)

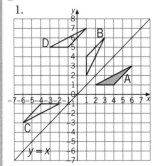

   b) Accurate diagrams to be drawn. Lengths to be accurate to ±2 mm, angles to ±2°.
   c) 130 km

Pages 58–59 Transformations
A
1. b
2. c
3. d

B
1.

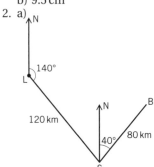

2. a) Reflection
   b) Translation
   c) Rotation
   d) Reflection

C

1.

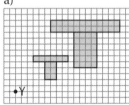

mirror line   mirror line

2.

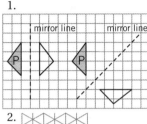

3. a)

   •Y

   b) 12.5 cm and 3.6 cm

Pages 60–61 Pythagoras' theorem

A

1. a) Card D
   b) Card F
   c) Card A
   d) Card E
   e) Card B
   f) Card C
2. $3^2 + 4^2 = 5^2$
   $9 + 16 = 25$
   Since Pythagoras' theorem works, the triangle must be right-angled.
3. a) RS = 7.8
   b) VW = 3.6
   c) XY = 4.5
4. 24.1 cm
5. 13.7 cm

B

1. 540.8 km
2. $17^2 = 15^2 + 8^2$
   $289 = 225 + 64$
   Since Pythagoras' theorem works, the triangle must be right-angled.
3. 15.1 cm
4. a) RP is 5 cm because using Pythagoras' theorem:
   $4^2 + 3^2 = 5^2$
   $16 + 9 = 25$
   b) 8.66 cm

Pages 62–63 Trigonometry

A

1. a) 23.6°
   b) 50.2°
   c) 31.9°
   d) 27.8°
   e) 34.8°

2. a) 6 cm
   b) 6.36 cm
   c) 11.49 cm
   d) 5.20 cm
   e) 13.86 cm
   f) 26.31 cm
3. 28.5°
4. 9.2 m

B

1. a) 9.8 km (1 d.p.)
   b) 057° (nearest degree)
2. 5.9° (1 d.p.)
3. 66.4° (1 d.p.)
4. 28.3 cm (1 d.p.)

Pages 64–65 Measures & measurement

A

1. c
2. b
3. a
4. c
5. d

B

1. a) 7000 m
   b) 0.6 m
   c) 0.5 m
   d) 2.5 kg
   e) 6000 kg
   f) 2600 ml
   g) 5.8 l
   h) 0.53 kg
2. Approx. 1.1 gallon
3. 44.4 mph
4. 1 hour 30 minutes
5. 22 miles
6. Lower limit 15.75
   Upper limit 15.85
7. $12.55 \leqslant h < 12.65$

C

1. 32 km
2. James has more sweets because 3 ounces is approximately 75 g.
3. 13 g/cm³
4. 98 miles
5. a) 25.5 metres
   b) 12.25 metres
   c) 54

Pages 66–67 Similarity

A

1. Rectangles A and C
2. a) 3.65 cm
   b) 4.98 cm
3. a) 6.1 cm
   b) 11.5 cm
   c) 10.7 cm
   d) 3.6 cm
4. a) True
   b) True

B

1. a) 30 cm
   b) Ratio of widths $15 : 25$
   $= 3 : 5$
   Capacity ratio $= 3^3 : 5^3$
   $= 27 : 125$

2. a) $y = 10$ cm
   b) $p = 8$ cm
   c) No – because they only have one common angle of 110°:
   $(110° + 30° + 40°)$
   other $(110° + 50° + 20°)$

Pages 68–69 2D shapes

A

1. a) $P = 12$ cm; $A = 6$ cm²
   b) $P = 20.57$ cm; $A = 25.13$ cm²
   c) $P = 25.13$ cm; $A = 50.27$ cm²
   d) $P = 46.8$ cm; $A = 61.09$ cm²
2. a) 122.5 cm²
   b) 63.6 cm²
3. a) 15.7 m
   b) 19.6 m²
4. 320 cm²

B

1. a) 4 cm
   b) 5 cm
2. a) 1256.6 cm²
   b) 1570.8 cm²
3. a) 21.935 cm²
   b) 0.002 193 5 m²

Pages 70–71 Volume of 3D solids

A

1. a
2. c
3. c
4. a

B

1. 1080 cm³
2. 682.5 cm³
3. 1.94 m³ or 1 940 000 cm³
4. 275 cm³
5. $\frac{4}{5}pq^2$, $\pi p^2 q$, $\frac{5}{6}pqr$

C

1. 8.1 m
2. a) 198 cm³
   b) 1584 cm³
3. $4\pi r^2$ is a formula for area. 4π is a number and has no effect. r^2 means length × length which gives area. You would need to multiply by another length to get volume.

HANDLING DATA
Pages 72–73 Collecting data

A

1. b
2. c
3. d

B

1.

Eye colour	Tally	Frequency

2. The tick boxes overlap; for example, in which box would somebody who watched 2 hours of television tick? Timescale? Is it per day? Per week? Need an extra box with 4 or more hours:

How much television do you watch per day?

less than 1 hr	1 up to 2 hrs	2 up to 3 hrs	3 up to 4 hrs	4+ hrs

```
5
4 | 2  8  7  2  4
5 | 4  3  0  6  4  9
6 | 1  7  3  9
7 | 3  1  1
    reordered =>
4 | 2  2  4  7  8
5 | 0  3  4  4  6  9
6 | 1  3  7  9
7 | 1  1  3
```

C

1.

Weight (w) in grams	Frequency
$0 \leqslant w < 5$	4
$5 \leqslant w < 10$	2
$10 \leqslant w < 15$	7
$15 \leqslant w < 20$	4
$20 \leqslant w < 25$	5
$25 \leqslant w < 30$	6

2. a) The middle three boxes overlap with each other, for example: age 15 could be placed in two boxes.
   b) less than 1 hour
   1 up to 2 hours
   2 up to 3 hours
   3 hours and over

Pages 74–75 Representing information

A

1.

Day	1	2	3	4	5	6	7	8
Rainfall in mm	12	4	7	2	5	1	2	6

2. a) 120°
   b) 8 hours
   c) 6 hours

B

1. a)

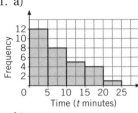

b)

Time (*t* minutes)	Frequency
$0 \le t < 5$	12
$5 \le t < 10$	8
$10 \le t < 15$	5
$15 \le t < 20$	4
$20 \le t < 25$	1

c) He is correct as two-thirds of the people queue for less than 10 minutes.

2. a)

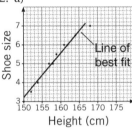

b) $\frac{8}{24} = \frac{1}{3}$

Pages 76–77 Scatter diagrams

A

1. a) Graph X
   b) Graph Z
   c) Graph Y
   d) Graph Y
   e) Graph Z

2. a)

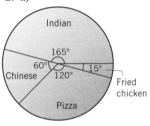

b) Positive correlation
c) See scatter diagram
d) 155.5 cm (approx)

B

1. The scale does not start at zero, so the growth looks much bigger than it actually is.
2. a) Positive correlation, the longer the hours spent revising the higher the maths score.
   b) Negative correlation, the longer the hours spent watching the television, the lower the maths score.

c) Zero or no correlation.
d) Approx. 76%

Pages 78–79 Averages 1

A

1. a) 3.2
   b) 2.5
   c) 2
   d) 6
2. 32 cm
3. a) 24.75
   b) Yes, as the mean is approximately 25.
4. She found the middle card but she did not put them in order of size first.

B

1. a) 8
   b) (i) 9
      (ii) 8
      (iii) 9
2. 7, 7, 13
3. a) 1.9
   b) mode = 2, median = 2.5

Pages 80–81 Averages 2

A

1. a
2. b
3. b
4. c

B

1. a) True
   b) False
   c) True
   d) True
   e) False
   f) False

C

1. a) 35 minutes
   b) $30 \le t < 40$
   c) 11.3 minutes
   d) $10 \le t < 15$
   e) The second set of students did the test in a shorter time, because the mean time was 11.3 minutes compared to 35 minutes for the first group.
2. Company B. The mean length was the same but the range was much smaller, hence the nails were probably more consistent in length.

Pages 82–83 Cumulative frequency

A

1. a) Median 'approx' 54
   b) Lower quartile 'approx' 40
   c) Upper quartile 'approx' 59
   d) Interquartile range 'approx' 19

2. a)

Time in minutes	Frequency	Cumulative frequency
$0 \le t < 10$	5	5
$10 \le t < 20$	20	25
$20 \le t < 30$	26	51
$30 \le t < 40$	18	69
$40 \le t < 50$	10	79
$50 \le t < 60$	4	83

b)

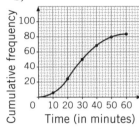

c) Approx. 26 minutes
d) Interquartile range ≈ 18 minutes
e) About 8 people

B

1. a) Approx. £21
   b) Approx. 68%
   c) Approx. £14
   d)

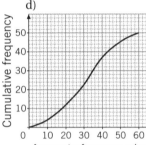

e) A ✓
   B ✗
   C ✗

Pages 84–85 Probability 1

A

1. a) $\frac{3}{13}$
   b) $\frac{6}{13}$
   c) 0
   d) $\frac{7}{13}$
2. a) $\frac{15}{28}$
   b) $\frac{13}{28}$
   c) 0
3. a) True
   b) False
   c) False
4. 140 people
5. 42 races

B

1. a) Spinner Y, because 3 occupies two of the spaces and not just one as on spinner X.
   b) Spinner X, because $\frac{3}{4}$ of the spinner has even numbers.
2. a) 0.15
   b) Tomato
   c) 0.4
   d) 58%
   e) 90 packets

Pages 86–87 Probability 2

A

1. a) $\frac{31}{150}$
   b) Approach the expected probability of $\frac{1}{6}$

2.

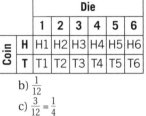

b) $\frac{1}{12}$
c) $\frac{3}{12} = \frac{1}{4}$
3. a)

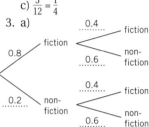

b) 0.32
c) 0.56

B

1. a) $\frac{7}{16}$
   b) $\frac{10}{16}$
2. a) 6 races
   b) 20 races
   c) (i) 0.36
      (ii) 0.48

ACKNOWLEDGEMENTS

The author and publisher are grateful to the
copyright holders for permission to use quoted
materials and images.

Every effort has been made to trace copyright
holders and obtain their permission for the use
of copyright material. The authors and
publishers will gladly receive information
enabling them to rectify any error or omission in
subsequent editions. All facts are correct at
time of going to press.

Published by Letts Educational Ltd.
An imprint of HarperCollinsPublishers
77–85 Fulham Palace Road
London W6 8JB

ISBN 9781844195121

First Published 2010

Text © Fiona C. Mapp 2007

Design and illustration ©2007 Letts Educational Ltd.

British Library Cataloguing in Publication Data.

A CIP record of this book is available from the
British Library.

Book Concept and Development: Helen Jacobs
Author: Fiona C. Mapp
Editorial: Marion Davies and Alan Worth
Cover Design: Angela English
Inside Concept Design: Starfish Design
Text Design, Layout and Editorial: MCS
Publishing Services

B

Answer all parts of the questions.

1 Here are four containers. Water is poured at a constant rate into three of the containers.

The graphs show the depth of the water as the containers fill up.

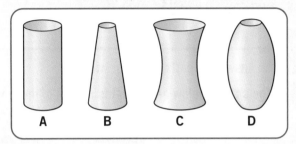

A   B   C   D

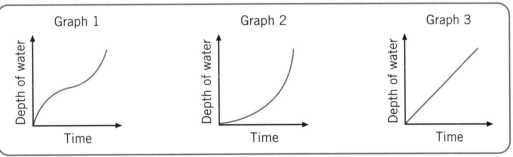

Graph 1   Graph 2   Graph 3

Fill in the gaps below to show which container matches each graph.

Graph 1 matches container ....   Graph 2 matches container ....

Graph 3 matches container ....   (3 marks)

2 The simplified graph shows the journey of a taxi travelling from London to St Albans.

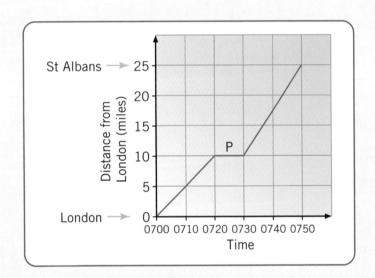

a) What is the taxi's average speed from London to St Albans? .... mph   (1 mark)

b) Explain what has happened at P.

.... (1 mark)

c) A different taxi travels from St Albans to London. It sets off at 0720 and drives at a constant speed directly to London, arriving at 0745.

On the graph, show the taxi's journey from St Albans to London.   (2 marks)

d) At approximately what time do the two taxis pass each other?

.... (1 mark)

Score   **/ 8**

Total score   **/ 16**

How well did you do?   ✗ 1–4 **Try again**   5–8 **Getting there**   9–12 **Good work**   13–16 **Excellent!** ✓

For more help on this topic see KS3 Maths 5–8 Success Guide pages 46–47.

Shapes

A

Answer all parts of all questions.

1 The shapes below are triangles. Which of these words best describes each of them:

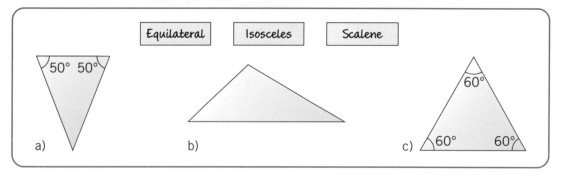

Equilateral   Isosceles   Scalene

a is ....   b is ....   c is .... (3 marks)

2 The lettered shapes are quadrilaterals. Decide whether each of the statements below is true or false.

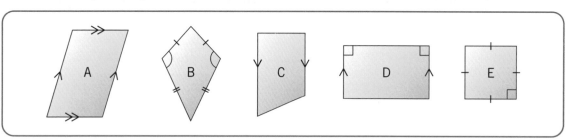

a) Shape A is a parallelogram. ....   b) Shape B is a kite. ....

c) Shape C is a parallelogram. ....   d) Shape D is a rectangle. ....

e) Shape E is a square. ....   (5 marks)

3 Complete the missing words in these statements.

a)   A trapezium has one pair of .... sides.

b)   A parallelogram has rotational symmetry of order ....

c)   A square has .... lines of symmetry.

d)   The perimeter of a circle is known as the ....

e)   The radius and tangent to a point make an angle of ....   (5 marks)

4 Draw a regular hexagon.   (1 mark)

a)   How many lines of symmetry does it have?... (1 mark)

b)   What is its order of rotational symmetry? .... (1 mark)

Score   / 16

B

Answer all parts of the questions.

1 a) Look at this shape. Two of the statements below are correct.
Tick the correct statements.

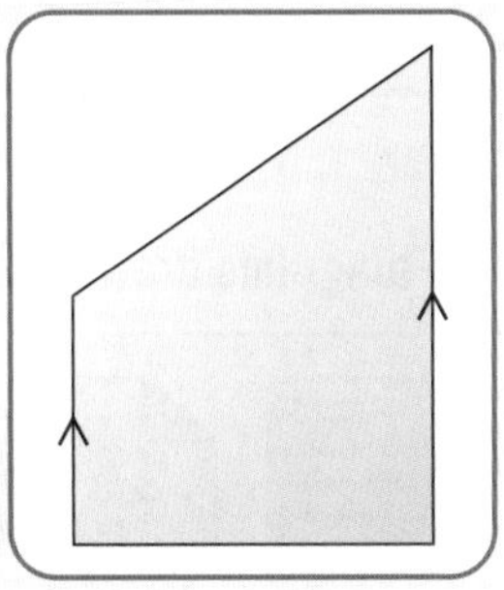

The shape is a hexagon. ☐

The shape is a quadrilateral. ☐

The shape is a kite. ☐

The shape is a trapezium. ☐

The shape is a parallelogram. ☐

(1 mark)

b) Look at this shape. Complete these statements about the shape.

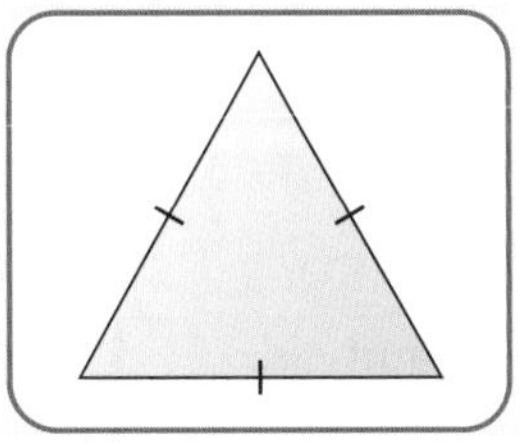

The shape is an .... triangle. The shape has .... lines of symmetry.

Each angle is .... degrees.

(3 marks)

2 Some names of polygons have been written on cards.

| Pentagon | Hexagon | Heptagon | Quadrilateral | Octagon |

Write down the name of the polygon for each of the shapes below, choosing from the list above.

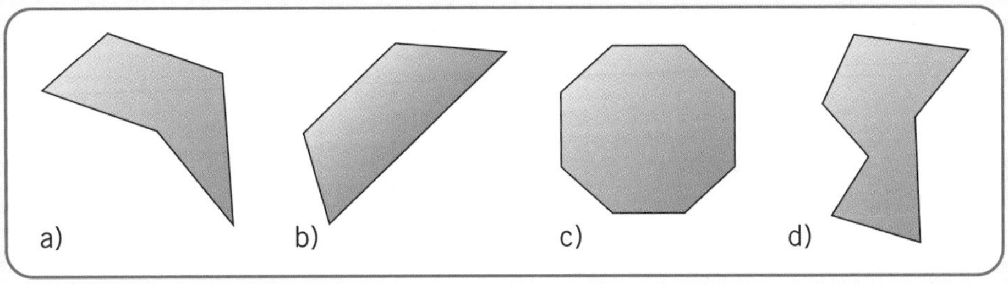

a)   b)   c)   d)

a) ....   b) ....

c) ....   d) ....   (4 marks)

3 In the spaces provided, draw an example of each of the shapes listed.

a) Circle   b) Pentagon

c) Parallelogram   d) Kite   (4 marks)

a)   b)   c)   d)

Score   / 12

Total score   / 28

How well did you do?   ✗ 1–7 Try again   8–13 Getting there   14–22 Good work   23–28 Excellent! ✓

For more help on this topic see KS3 Maths 5–8 Success Guide pages 50–51.

Solids

A

Choose just one answer, a, b, c or d.

Questions 1–5 relate to the diagrams drawn below.

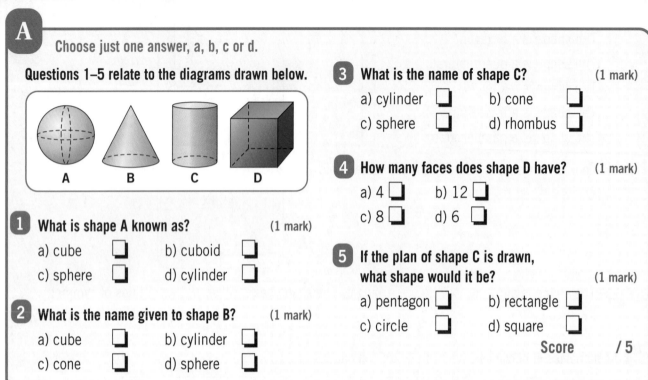

A   B   C   D

1 What is shape A known as?   (1 mark)

a) cube ☐   b) cuboid ☐

c) sphere ☐   d) cylinder ☐

2 What is the name given to shape B?   (1 mark)

a) cube ☐   b) cylinder ☐

c) cone ☐   d) sphere ☐

3 What is the name of shape C?   (1 mark)

a) cylinder ☐   b) cone ☐

c) sphere ☐   d) rhombus ☐

4 How many faces does shape D have?   (1 mark)

a) 4 ☐   b) 12 ☐

c) 8 ☐   d) 6 ☐

5 If the plan of shape C is drawn,
what shape would it be?   (1 mark)

a) pentagon ☐   b) rectangle ☐

c) circle ☐   d) square ☐

Score   / 5

B

Answer all parts of all questions.

1 Which of the following nets would make a cube? .... (1 mark)

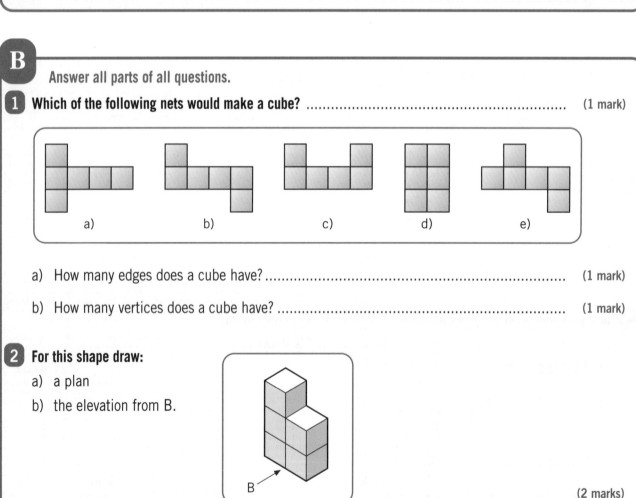

a)   b)   c)   d)   e)

a) How many edges does a cube have? .... (1 mark)

b) How many vertices does a cube have? .... (1 mark)

2 For this shape draw:

a) a plan

b) the elevation from B.

B

(2 marks)

Score   / 5

Answer all parts of the questions.

1 Jeremy is making a box to hold a small chocolate egg.

He decides to draw a sketch of the net of the box.

The base of the box is shaded.

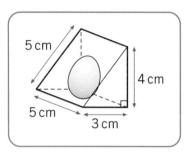

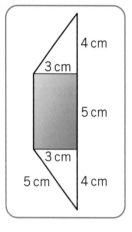

a) Complete the sketch of the net of the box. (1 mark)

b) Jacqueline is also making a box to hold a small chocolate egg.
   Her box has no lid.
   On a separate piece of paper, accurately draw a net for the box.

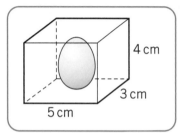

(3 marks)

2 This diagram shows a model made from 9 cubes. The cubes are either mauve or white. There are five mauve cubes and four white cubes.

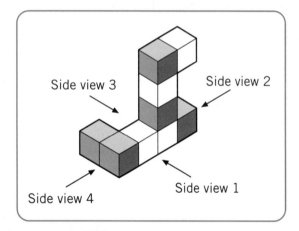

Side view 3   Side view 2

Side view 4   Side view 1

a) The drawings below show the side views of the model.
   Write the numbers to show which side view each drawing represents.

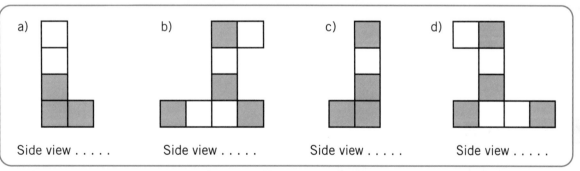

a)   b)   c)   d)

Side view . . . . .   Side view . . . . .   Side view . . . . .   Side view . . . . .

(2 marks)

b) On a separate piece of paper, draw the top view of the model, (2 marks)
   making sure you shade the mauve cubes.

Score   / 8

Total score   / 18

How well did you do?   ✗ 1–4 **Try again**   5–8 **Getting there**   9–13 **Good work**   14–18 **Excellent!**   ✓

For more help on this topic see KS3 Maths 5–8 Success Guide pages 52–53.

Constructions & LOGO

A

Answer all parts of all questions.

1 Here is a computer program to draw a triangle:

Forward 6

Turn right 120°

Forward..................................

....

....

Complete the instructions to draw this shape.

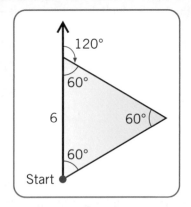

(1 mark)

2 Use a ruler and compasses to draw an accurate diagram of each of the following shapes on a separate piece of paper.

a)

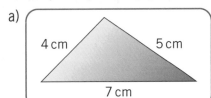

4 cm 5 cm

7 cm

(2 marks)

b)

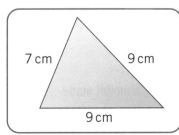

7 cm 9 cm

9 cm

(2 marks)

3 Use a ruler and a protractor to draw the following shapes on a separate piece of paper.

a)

6 cm

80°

7 cm

(1 mark)

b)

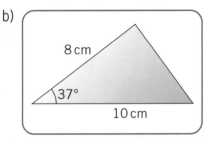

8 cm

37°

10 cm

(1 mark)

c)

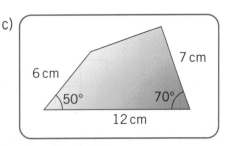

7 cm

6 cm

50° 70°

12 cm

(1 mark)

4 Construct the perpendicular bisector of this line AB.

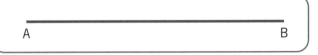

A   B

(2 marks)

5 Using ruler and compasses only, bisect this angle.

(2 marks)

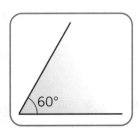

60°

Score   / 12

50

Answer all parts of the questions.

1 **Shape A is a square.**

The instructions to draw shape A are:

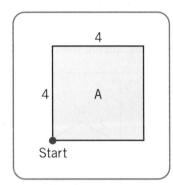

Forward 4
Turn right 90°
Forward 4
Turn right 90°
Forward 4
Turn right 90°
Forward 4

a) Write down the instructions to draw a square that has sides three times the length of those of shape A.

....

....

.... (2 marks)

b) Shape B is a parallelogram.

Complete the instructions to draw shape B.

Forward 5

Turn right 60°

....

....

....

....

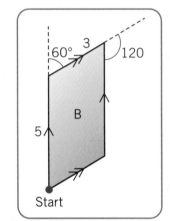

(2 marks)

2 **Joshua has made a rough sketch of this triangle.**

Using compasses, construct the triangle accurately on a separate piece of paper.

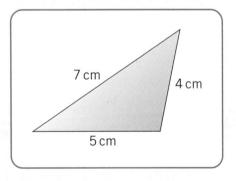

(2 marks)

Score   / 6

Total score   / 18

How well did you do?   ✗ 1–4 **Try again**   5–8 **Getting there**   9–12 **Good work**   13–18 **Excellent!**   ✓

For more help on this topic see KS3 Maths 5–8 Success Guide page 54.

Loci & coordinates in 3D

A

Answer all parts of all questions.

1 **For each of the statements below, decide whether it is true or false.**

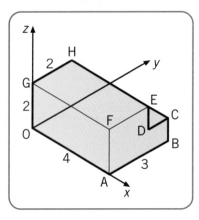

a) The coordinates of point A are (4, 0, 0)... (1 mark)

b) The coordinates of point H are (0, 2, 2)... (1 mark)

c) The coordinates of point E are (2, 4, 2)... (1 mark)

d) The coordinates of point C are (4, 3, 1)... (1 mark)

2 **Tess buries her bone in the garden. The garden is surrounded by a fence ABCD, as shown.**

i) The bone is buried at least 2 metres from the greenhouse.

ii) The bone is buried at least 3 metres from the centre of the pond (P).

iii) The bone is buried at least 1 metre from the house.

Shade in all the possible areas where Tess could have buried her bone.

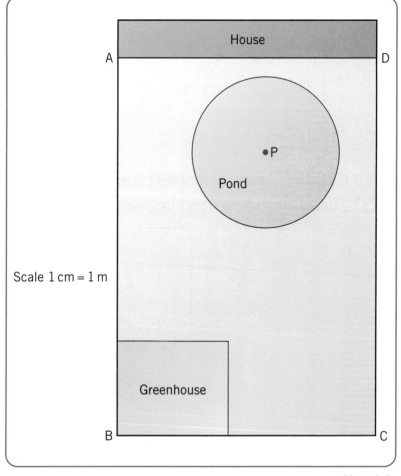

(4 marks)

Score   / 8

B

Answer all parts of the questions.

1 The diagram shows the positions of three towns, C, D and E.

A supermarket (S) is to be built so that it is:

i) equidistant from D and E

ii) 12 km from town C.

By drawing the loci of i) and ii) and using a scale of 1 cm = 4 km, clearly mark the position of the supermarket.

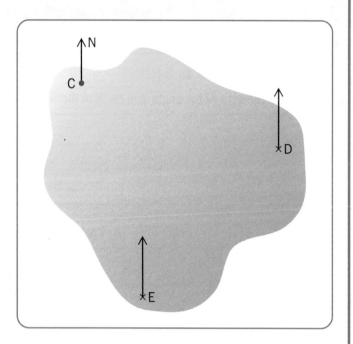

(3 marks)

2 In the scale drawing opposite, the shaded area represents a fish pond. There is a wire fence all around the pond. The shortest distance from the fence to the edge of the pond is always 2 m.

On the diagram, draw accurately the position of the fence.

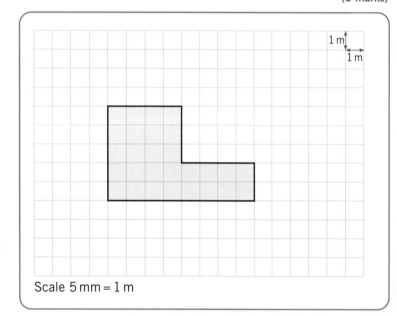

Scale 5 mm = 1 m

(3 marks)

3 The diagram shows a solid. Complete the coordinates for each of the vertices listed below.

R = (................, ..., ...)

S = (................, ..., ...)

T = (................, ..., ...)

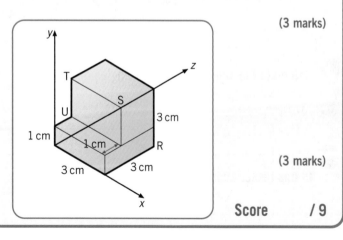

(3 marks)

Score   / 9

Total score   / 17

How well did you do?   ✗ 1–3 Try again   4–8 Getting there   9–13 Good work   14–17 Excellent!   ✓

For more help on this topic see KS3 Maths 5–8 Success Guide page 55.

Angles & tessellations

A

Answer all parts of all questions.

1 Work out the size of the angle *a* in each of the diagrams.

a)
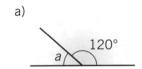

b)

c)

a = ...

a = ...

a = ...

d)

e)

f)

a = ...

a = ...

a = ... (6 marks)

2 Calculate the size of the angles marked with letters.

a)
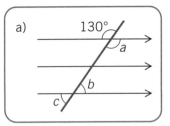

b)
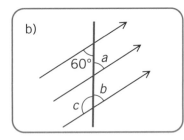

a) *a* = ... *b* = ... *c* = ...

b) *a* = ... *b* = ... *c* = ... (6 marks)

3 The size of an exterior angle of a regular polygon is 18°.
How many sides does the polygon have?

.... (1 mark)

4 Calculate the sum of the interior angles of a hexagon.

Ⓒ .... (1 mark)

5 Add six more shapes
to this tessellation.

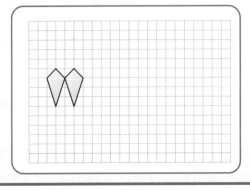

(2 marks)

Score   / 16

Ⓒ *Indicates that a calculator may be used*

Answer all parts of the questions.

1 a) Fiona has drawn a shape
using her computer.

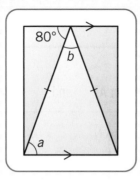

The diagram shows an isosceles triangle inside a rectangle.

i) Calculate the size of the angle marked *a*.   *a* = ... °   (1 mark)

ii) Calculate the size of the angle marked *b*.   *b* = ... °   (1 mark)

b) Simon has drawn this shape
with his computer.

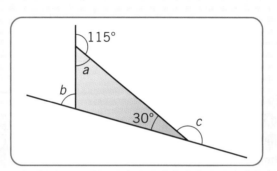

Calculate angles *a*, *b* and *c*.
Show your working.

a = ...   *b* = ...   *c* = ...   (3 marks)

c) Simon draws a second shape
with his computer.

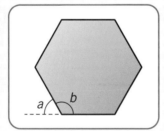

It is a regular hexagon.

i) Calculate the sizes of angles *a* and *b*.
Show your working.

a = ...   *b* = ...   (2 marks)

ii) Explain why a regular hexagon will tessellate.

....

....   (1 mark)

Score   / 8

Total score   / 24

Bearings & scale drawings

A

Answer all parts of all questions.

1 Calculate the three-figure bearing of: i) B from A   and   ii) A from B in each diagram.

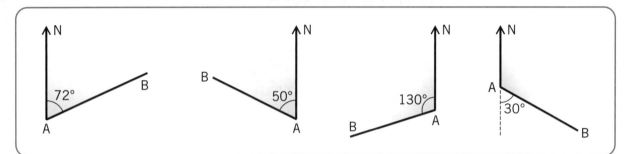

a) ....   b) ....   c) ....   d) ....

....

(8 marks)

2 Preachers Bottom is 8 km due north of Hookfield. Marigold Hill is on a bearing of 060° from Hookfield, whilst it is on a bearing of 110° from Preachers Bottom.

Make a careful scale drawing, using a scale of 1 cm to 1 km and find the distance of Marigold Hill from Hookfield.

Measure the bearing of Hookfield from Marigold Hill.

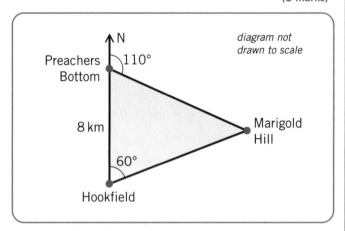

(3 marks)

3 The scale on a road map is 1 : 50000. If two towns are 20 cm apart on the map, work out the real distance, in kilometres, between the two towns.

C .... (2 marks)

Score   / 13

C *Indicates that a calculator may be used*

Answer all parts of the questions.

1 Yen is making a stencil.
He makes a sketch of the
stencil on some paper.

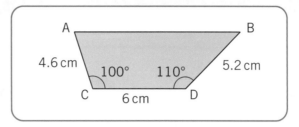

a) Make an accurate full-size drawing of the stencil in the space provided.

(4 marks)

b) Measure the length of AB to the nearest 0.1 cm. .... cm (1 mark)

2 a) The diagram shows the position of a lighthouse (L). A ship (S) is 120 km from the lighthouse on a
bearing of 140°. If 1 cm = 20 km, mark accurately on the diagram the position of the ship.

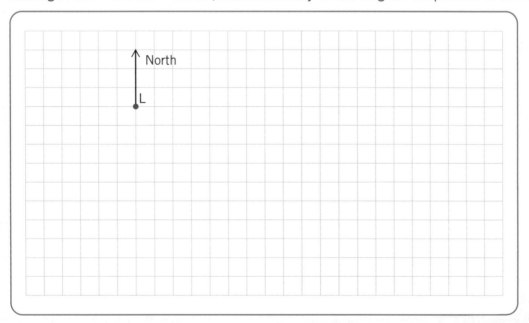

(3 marks)

b) A buoy (B) is 80 km and on a bearing of 040° from the ship.
Mark accurately on the diagram the position of the buoy. (2 marks)

c) How far is the lighthouse from the buoy? .... km (1 mark)

Score   / 11

Total score   / 24

How well did you do?   ✗ 1–7 Try again   8–11 Getting there   12–17 Good work   18–24 Excellent! ✓

For more help on this topic see KS3 Maths 5–8 Success Guide pages 58–59.

Transformations

A

Choose just one answer, a, b, c or d.

Questions 1–3 refer to the diagram below.

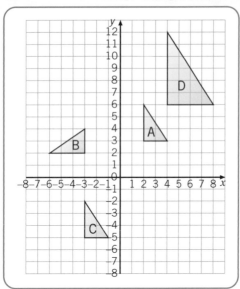

1 What is the transformation that would map shape A onto shape B? *(1 mark)*

a) reflection ☐ b) rotation ☐
c) translation ☐ d) enlargement ☐

2 What is the transformation that would map shape A onto shape C? *(1 mark)*

a) reflection ☐ b) rotation ☐
c) translation ☐ d) enlargement ☐

3 What is the transformation that would map shape A onto shape D? *(1 mark)*

a) reflection ☐ b) rotation ☐
c) translation ☐ d) enlargement ☐

Score   / 3

B

Answer all parts of all questions.

1 On the grid opposite, draw the image of the shaded shape after the following transformations:

a) a reflection in the line $y = x$;
   call the image B

b) a rotation of 180° about the origin (0, 0);
   call the image C

c) a translation by the vector $\begin{pmatrix} -5 \\ 4 \end{pmatrix}$;
   call the image D.

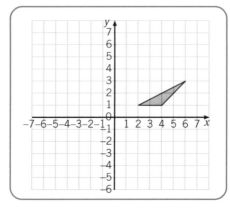

(3 marks)

2 Each of the following shapes is the result of a translation, a reflection, or a rotation of object A. State the transformation that has taken place for each of the following:

a) A is transformed to B

.... (1 mark)

b) A is transformed to C

.... (1 mark)

c) A is transformed to D

.... (1 mark)

d) A is transformed to E

.... (1 mark)

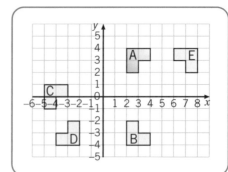

Score   / 7

Answer all parts of the questions.

1 Reflect triangle P in the mirror line on each of these diagrams.

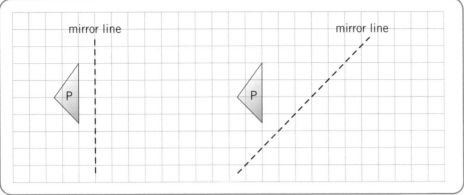

(2 marks)

2 Five cubes join together to make a ⬡ shape.
The diagram shows the ⬡ shape after quarter turns in one direction.
On the paper below, draw the ⬡ shape after the next quarter turn in the same direction.

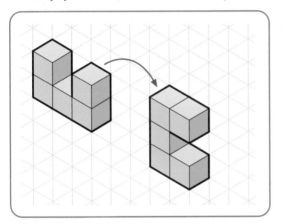

(2 marks)

a) The grid shows a T shape. On the grid draw an enlargement of scale factor 2 of the T shape. Use point Y as the centre of enlargement.

(2 marks)

b) This sketch shows two rectangles. The bigger rectangle is an enlargement of scale factor 2.5 of the smaller rectangle. Write down the two missing values.

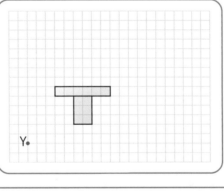

5 cm

. . . . .

. . . . .

9 cm

diagrams not drawn to scale

(2 marks)

Score   / 8

Total score   / 18

How well did you do? ✗ 1–3 **Try again**   4–8 **Getting there**   9–13 **Good work**   14–18 **Excellent!** ✓

For more help on this topic see KS3 Maths 5–8 Success Guide pages 60–61.

Pythagoras' theorem

A

Answer all parts of all questions.

1 Giovanna is playing a game. She writes the missing lengths of some triangles onto cards.

| 16.77 | 17 | 17.69 | 15 | 17.86 | 15.65 |
| A | B | C | D | E | F |

Match each length with the correct triangle. **(6 marks)**

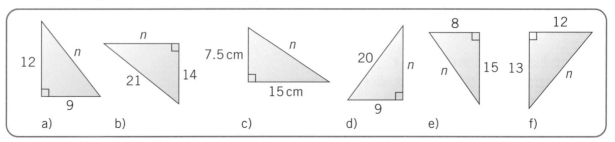

a) b) c) d) e) f)

2 Hannah says:

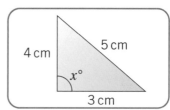

Angle x must be 90°.

(triangle with sides 4 cm, 5 cm, 3 cm, angle $x°$)

Explain how Hannah knows this without measuring the size of the angle.

....

.... **(2 marks)**

3 Calculate the distances of these lines:

a) RS ....

b) VW ....

c) XY ....

(3 marks)

4 Calculate the length of the diagonal of this rectangle. **(2 marks)**
Give your answer to 1 decimal place.

(rectangle 22.3 cm by 9.2 cm)

....

5 Calculate the perpendicular height of this isosceles triangle. **(2 marks)**
Give your answer to 1 decimal place.

(triangle with two sides 15 m and base 12 m)

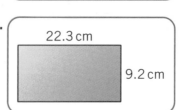

....

Score   / 15

C *Indicates that a calculator may be used*

Answer all parts of the questions.

1 A ship sails from Miami to an island in the Atlantic. The island is 300km North of Miami and
C 450km East of Miami.

Calculate the shortest distance
between the island and Miami.

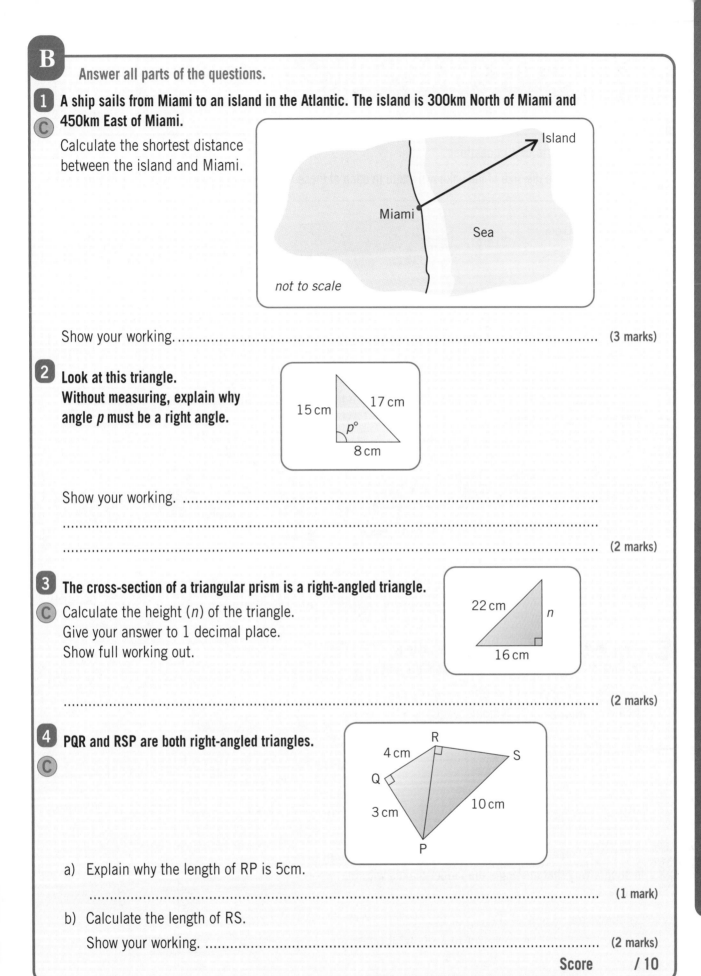

not to scale

Show your working. .... (3 marks)

2 Look at this triangle.
Without measuring, explain why
angle *p* must be a right angle.

15 cm   17 cm

p°

8 cm

Show your working. ....
....
.... (2 marks)

3 The cross-section of a triangular prism is a right-angled triangle.
C Calculate the height (*n*) of the triangle.
Give your answer to 1 decimal place.
Show full working out.

22 cm   *n*

16 cm

.... (2 marks)

4 PQR and RSP are both right-angled triangles.
C

R
4 cm   S
Q
3 cm   10 cm
P

a) Explain why the length of RP is 5cm.

.... (1 mark)

b) Calculate the length of RS.

Show your working. .... (2 marks)

Score / 10

Total score / 25

PYTHAGORAS' THEOREM   Shape, space and measures

How well did you do?   ✗ 1–6 Try again   7–11 Getting there   12–18 Good work   19–25 Excellent! ✓

For more help on this topic see KS3 Maths 5–8 Success Guide pages 62–63.

61

Trigonometry

A

Answer all parts of all questions.

1 **LEVEL 8** Calculate the size of the unknown angle in each of these triangles. Give your answers to 1 decimal place.

Ⓒ

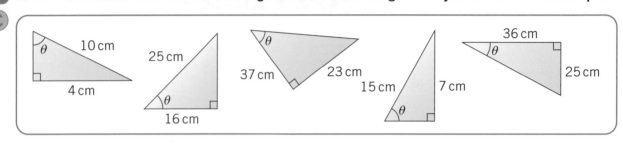

a) .... b) .... c) .... d) .... e) .... (5 marks)

2 **LEVEL 8** Calculate the length of side x in each of the triangles below. Give your answers to 2 decimal

Ⓒ places, where necessary.

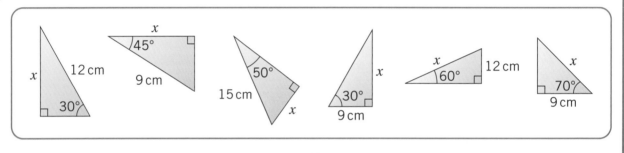

a) .... b) .... c) .... d) .... e) .... f) .... (6 marks)

3 **LEVEL 8** A tower of height 25 metres casts a

Ⓒ shadow of length 46 metres on horizontal ground.

Calculate the angle of elevation of the sun.
Give your answer to 1 decimal place.

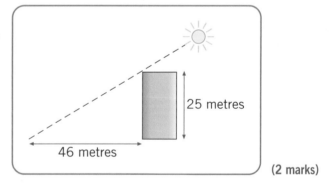

.... (2 marks)

4 **LEVEL 8** A ladder of length 12 metres rests against a wall

Ⓒ in such a way that the angle the ladder makes with the
wall is 40°.

Calculate the height of the top of the ladder above
the ground, giving your answer to 1 decimal place.

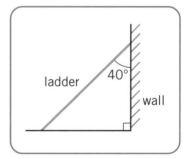

.... (2 marks)

Score   / 15

Ⓒ *Indicates that a calculator may be used*

B

Answer all parts of the questions.

1 **LEVEL 8** **Eagle Point is 8.2 km East**
C **and 5.4 km North of Briarton.**

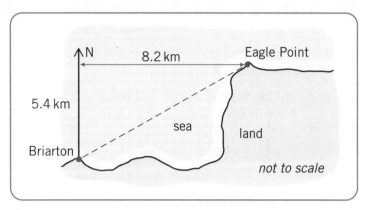

not to scale

a) Calculate the direct distance from Briarton to Eagle Point.

Show your working. .... (2 marks)

b) Ellen wants to sail directly from Briarton to Eagle Point.
On what bearing should she sail?

Show your working. .... (2 marks)

2 **LEVEL 8** **DEF and DFG are both**
C **right-angled triangles:**

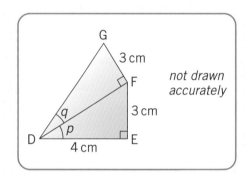

not drawn
accurately

By how many degrees is angle p
bigger than angle q?

Show your working. .... degrees   (4 marks)

3 **LEVEL 8** **Calculate the size of angle m**
C **in this isosceles triangle.**

Show your working.

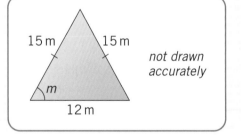

not drawn
accurately

m = ...

....   (2 marks)

4 **LEVEL 8** **Calculate the perimeter of this triangle.**
C **Give your answer to 1 decimal place.**

Show your working.

perimeter = ...

....

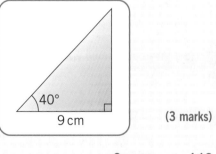

(3 marks)

Score   / 13

Total score   / 28

How well did you do?   ✗ 1–6 Try again   7–11 Getting there   12–19 Good work   20–28 Excellent! ✓

For more help on this topic see KS3 Maths 5–8 Success Guide pages 64–65.

63

TRIGONOMETRY   Shape, space and measures

Measures & measurement

A

Choose just one answer, a, b, c or d.

1 How many millimetres are in 5.2 cm? (1 mark)

a) 0.52 mm ☐   b) 5200 mm ☐

c) 52 mm ☐   d) 520 mm ☐

2 Approximately how many pounds are in 3 kg? (1 mark)

a) 4.5 ☐   b) 6.6 ☐

c) 5.8 ☐   d) 9 ☐

3 Nigel is 175 cm tall. What is the lower limit of his height? (1 mark)

a) 174.5 cm ☐   b) 175 cm ☐

c) 175.5 cm ☐   d) 174.9 cm ☐

4 A car travels 150 km in 1 hour and 20 minutes. What is the average speed of the car in km per hour? (1 mark)

a) 100 ☐   b) 125 ☐

c) 112.5 ☐   d) 120 ☐

5 What is the volume of a piece of wood with a density of 680 kg/m^3 and a mass of 34 kg? (1 mark)

a) 0.5 m^3 ☐   b) 20 m^3 ☐

c) 2 m^3 ☐   d) 0.05 m^3 ☐

Score   / 5

B

Answer all parts of all questions.

1 Complete the statements below.

a) 7 km = ... m   b) 60 cm = ... m

c) 500 mm = ... m   d) 2500 g = ... kg

e) 6 tonnes = ... kg   f) 2.6 l = ... ml

g) 580 cl = ... l   h) 530 g = ... kg   (8 marks)

2 A container holds 5 litres of water.

Ⓒ Approximately how many gallons is this? .... (1 mark)

3 Joanne drove 200 miles in 4 hours and 30 minutes.

Ⓒ At what average speed did she travel? .... (2 marks)

4 Tracey drove 90 miles at an average speed of 60mph.

Ⓒ How long did it take her? .... (2 marks)

5 How far does Ahmed walk if he walks at a speed of 4 miles per hour for 5 hours and 30 minutes?

Ⓒ .... (2 marks)

6 Thomas runs a race. His time is 15.8 seconds to the nearest tenth of a second.

Write down the upper and lower limits of his time. .... (2 marks)

7 The height (*h*) of a plant is 12.6 cm to the nearest tenth of a centimetre. Complete this inequality which shows the values between which the height of the plant can lie.

$12.55 \leqslant h < 12.$ .... (1 mark)

Score   / 18

Ⓒ *Indicates that a calculator may be used*

C

Answer all parts of the questions.

1 **How many kilometres are there in 20 miles?**

Complete the missing part of this sign.

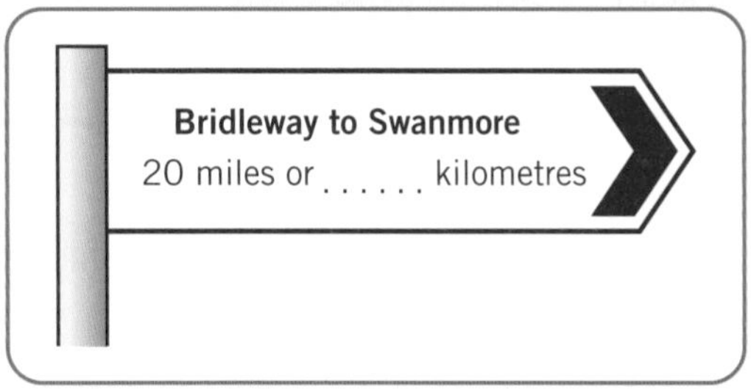

Bridleway to Swanmore
20 miles or . . . . . . kilometres

(1 mark)

2 **Jessica and James have some sweets.**

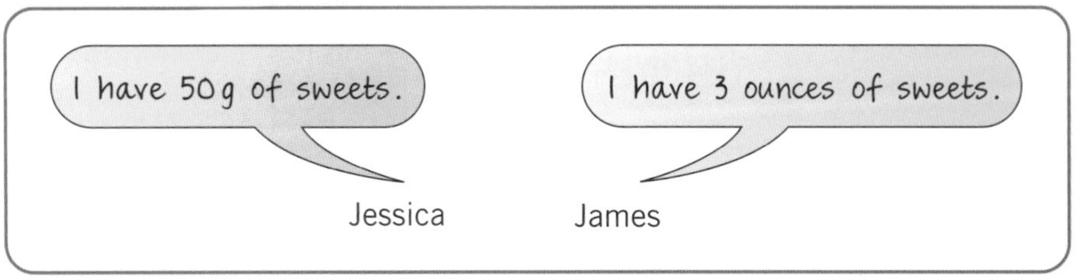

I have 50 g of sweets.

I have 3 ounces of sweets.

Jessica   James

Who has more sweets?

Explain your answer. .... (2 marks)

3 **The mass of a toy is 650 g. If the volume of the toy is 50 cm³, work out the toy's density.**

C .... (2 marks)

4 **A bus travels for $3\frac{1}{2}$ hours at an average speed of 28 mph. How far does the bus travel?**

.... (1 mark)

5 **A flower garden has been marked out as shown in the diagram.**

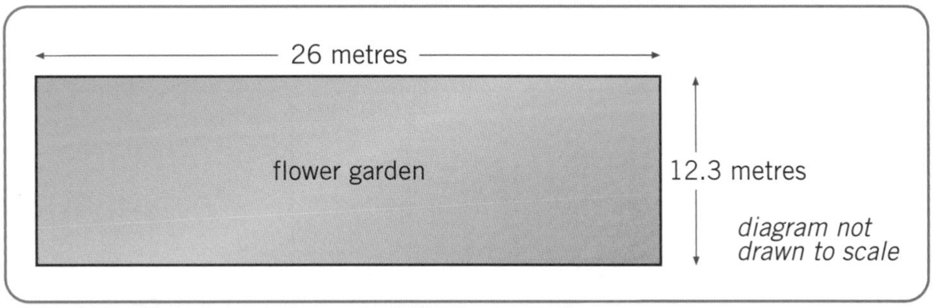

26 metres

flower garden   12.3 metres

diagram not drawn to scale

a) The length of the flower garden is 26 metres, to the nearest metre.
    What is the shortest possible length of the flower garden?

    .... m (1 mark)

b) The width of the flower garden is 12.3 metres to the nearest tenth of a metre.
    What is the shortest possible width of the flower garden?

    .... (1 mark)

c) Fifty rose bushes (to the nearest ten) are planted in the garden.
    What is the maximum number of rose bushes that are planted?

    .... (1 mark)

Score   / 9

Total score   / 32

How well did you do?   ✗ 1–7 **Try again**   8–15 **Getting there**   16–24 **Good work**   25–32 **Excellent!** ✓

For more help on this topic see KS3 Maths 5–8 Success Guide pages 66–68.

Similarity

A

Answer all parts of all questions.

1 **LEVEL 8** **Here are three rectangles. Which two are similar?**

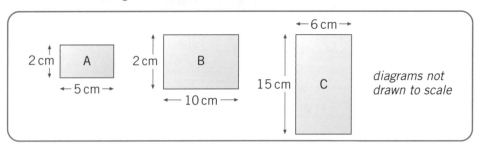

2 cm | A | ←5 cm→

2 cm | B | ←10 cm→

←6 cm→ | 15 cm | C

diagrams not drawn to scale

.... **(1 mark)**

2 **LEVEL 8** Triangle PQR is
C similar to triangle MNO.
Angle PQR = Angle MNO
Angle PRQ = Angle MON

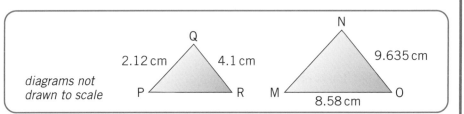

2.12 cm Q 4.1 cm

diagrams not drawn to scale

P R

N 9.635 cm

M 8.58 cm O

a) Calculate the length of PR, giving your answer to 3 s.f. ....cm **(2 marks)**

b) Calculate the length of MN, giving your answer to 3 s.f. ....cm **(2 marks)**

3 **LEVEL 8** Calculate the lengths marked x in these similar shapes.
C Give your answers correct to 1 decimal place.

a)
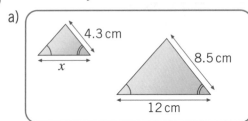

4.3 cm
x
8.5 cm
12 cm

b)
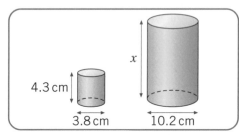

x
4.3 cm
3.8 cm
10.2 cm

$x =$ .... $x =$ ....

c)
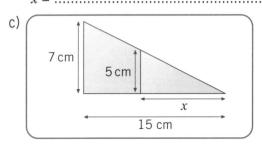

7 cm
5 cm
x
15 cm

d)
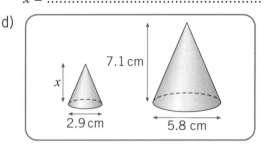

7.1 cm
x
2.9 cm
5.8 cm

$x =$ .... $x =$ .... **(8 marks)**

4 **LEVEL 8** Decide whether these statements are true or false.

a) If two shapes are similar, corresponding sides are in the same ratio. ....

b) If two shapes are similar, corresponding angles are the same. .... **(2 marks)**

Score   / 15

 Indicates that a calculator may be used

B

Answer all parts of the questions.

1 **LEVEL 8** **These vases are mathematically similar.**
The dimensions are shown.

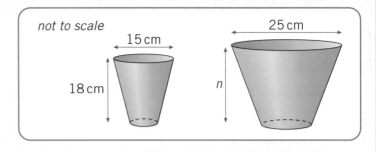

a) Calculate the value of *n*.

Show your working.

n = ... cm (2 marks)

b) Think about the ratio of the widths of the two vases.
Explain why the ratio of the capacity of the smaller pot to the capacity of the larger pot is 27 : 125.

.... (1 mark)

2 **LEVEL 8** **These triangles are similar.**

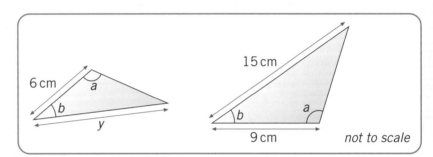

a) What is the value of *y*?

Show your working.

y = ... cm (2 marks)

b) Triangles ABC and BDE are similar.
What is the value of *p*?

Show your working.

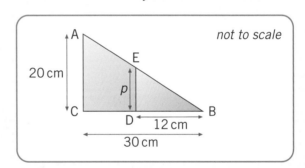

p = ... cm (2 marks)

c) Look at these triangles.
Are they similar?

Show your working.

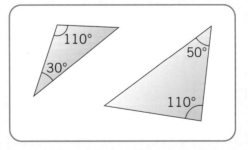

.... (1 mark)

Score /8

Total score /23

How well did you do? ✗ 1–6 **Try again** 7–11 **Getting there** 12–16 **Good work** 17–23 **Excellent!** ✓

For more help on this topic see KS3 Maths 5–8 Success Guide page 69.

Area & perimeter of 2D shapes

A

Answer all parts of all questions.

1 For each of these shapes, calculate the perimeter (P) and the area (A).

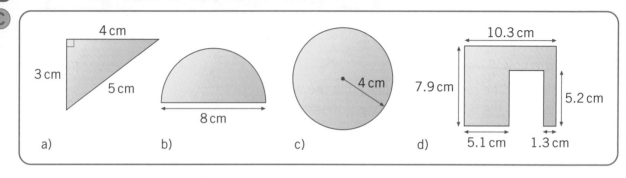

a)   b)   c)   d)

a)......................... b)......................... c)......................... d) .... 8 marks)

2 Work out the areas of these shapes. Give your answers to 1 decimal place.

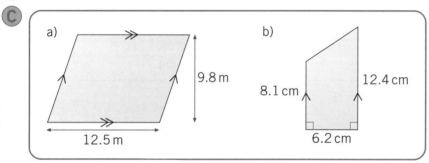

a).. b).. (4 marks)

3 A circular rose garden has a diameter of 5 m.

Find:

a) the circumference of the garden .... (1 mark)

b) the area of the garden.. (2 marks)

4 The area of a shape is 20 cm². The lengths of the shape are enlarged by a scale factor of 4. What is the area of the enlarged shape?

.... (1 mark)

Score   / 16

 Indicates that a calculator may be used

B

Answer all parts of the questions.

1 Each shape in this question has an area of 20 cm².

a) Calculate the height of this rectangle.

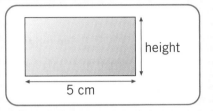

area = 20 cm²

height = ... cm   (1 mark)

b) Calculate the height of this triangle.

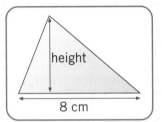

area = 20 cm²

height = ... cm   (1 mark)

2 a) A circle has a radius of 20 cm.

Calculate the area of the circle. Show your working.   ....cm²   (2 marks)

b) A circle of radius 20 cm sits inside a circle of radius 30 cm. Calculate the area of the shaded region.

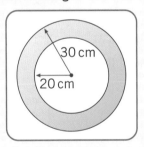

Show your working.   ....cm²   (3 marks)

3 A sewing pattern is in the shape of a trapezium.

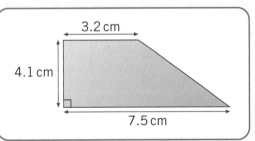

a) Calculate the area of the trapezium.
Show your working. .... cm²   (2 marks)

b) What is the area of the trapezium in square metres? .... m²   (1 mark)

Score   / 10

Total score   / 26

How well did you do?   1–6 Try again   7–11 Getting there   12–17 Good work   18–26 Excellent!

For more help on this topic see KS3 Maths 5–8 Success Guide pages 70–71.

Volume of 3D solids

A

Choose just one answer, a, b, c or d.

1 The volume of a cuboid is 20 cm³.
If the height is 1 cm and the width is 4 cm,
what is the length? **(1 mark)**

a) 5 cm ☐   b) 10 cm ☐

c) 15 cm ☐   d) 8 cm ☐

2 A cube of volume 2 cm³ is enlarged by a
scale factor of 3.
What is the volume of the enlarged cube? **(1 mark)**

a) 6 cm ☐   b) 27 cm³ ☐

c) 54 cm³ ☐   d) 18 cm ☐

3 LEVEL 8 The letter b represents a length.
What does this formula represent? $\dfrac{4b^3}{\pi}$ **(1 mark)**

a) length ☐   b) area ☐

c) volume ☐   d) nothing ☐

4 LEVEL 8 The letters a and b represent lengths.
What does this formula represent? $\sqrt{a^2 - b^2}$
(1 mark)

a) length ☐   b) area ☐

c) volume ☐   d) nothing ☐

Score   / 4

B

Answer all parts of all questions.

1 The diagram shows a triangular prism.
(C) Work out the volume of the triangular prism,
clearly stating your units.
Give your answer to 3 significant figures.

....

....

diagrams not drawn to scale
9 cm — 0.2 m — 12 cm

(2 marks)

2 The shape opposite is the cross-section of a
(C) prism, 15 cm long. Find the volume of the prism,
clearly stating your units.

....

....

5 cm — 6.5 cm — 9 cm

(2 marks)

3 Calculate the volume of the oil drum,
(C) clearly stating your units.
Give your answer to 3 significant figures.

....

....

125 cm — EXPRESS OIL — 1.58 m

(2 marks)

4 A child's toy is made up of a cuboid and a
(C) cylinder. Calculate the volume of the toy, giving
your answer to 3 significant figures.

....

....

3 cm — 5 cm — 8 cm — 5 cm — 6 cm

(2 marks)

(C) *Indicates that a calculator may be used*

5 **LEVEL 8 Here are some expressions.**

$4\pi r^2$	$\frac{4}{5}pq^2$	$\sqrt{p^2 + q^2}$	$\pi p^2 q$	$\frac{5}{6}pqr$	$\frac{\pi p^2}{q}$

The letters p, q and r represent lengths.
π, 4, 5, 6 are numbers that have no dimensions.
Three of the expressions represent volume. Tick the boxes underneath
these expressions.

(3 marks)

Score **/ 11**

C

Answer all parts of the questions.

1 **Some plates are put into a box. There are two types of box. If the volume of each box is the same, calculate the missing height of box B.**

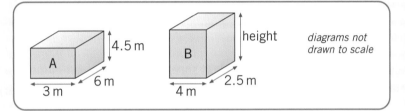

4.5 m

A

6 m

3 m

height

B

2.5 m

4 m

diagrams not drawn to scale

.... cm   (3 marks)

2 **A piece of cheese is in the shape of a triangular prism.**
C

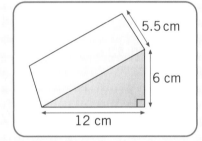

5.5 cm

6 cm

12 cm

a) Calculate the volume of the piece of cheese.
   Show your working. .... cm^3 (2 marks)

b) A larger block of cheese is exactly the same shape but each of the lengths is double that of the
   first piece of cheese.
   Calculate the volume of the larger block of cheese.
   Show your working. .... cm^3 (2 marks)

3 **LEVEL 8 Explain why the formula $4\pi r^2$, where r is a length, cannot be a formula for volume.**

....   (1 mark)

Score **/ 8**

Total score **/ 23**

How well did you do?   ✗ 1–4 Try again   5–10 Getting there   11–16 Good work   17–23 Excellent! ✓

For more help on this topic see KS3 Maths 5–8 Success Guide pages 72–73.

71

Collecting data

A

Choose just one answer, a, b, c or d.

Use the frequency table given to answer the next three questions.

Mass (*M* kg)	Frequency
$40 \leqslant M < 50$	4
$50 \leqslant M < 60$	10
$60 \leqslant M < 70$	15
$70 \leqslant M < 80$	25
$80 \leqslant M < 90$	4

1 How many people had a mass between 50 and 60 kg? **(1 mark)**

a) 4 ☐   b) 10 ☐   c) 15 ☐   d) 25 ☐

2 How many people in total were surveyed about their mass? **(1 mark)**

a) 55 ☐   b) 30 ☐   c) 58 ☐   d) 60 ☐

3 If somebody has a mass of 70 kg, into which class interval would they go? **(1 mark)**

a) $40 \leqslant M < 50$ ☐   b) $50 \leqslant M < 60$ ☐

c) $60 \leqslant M < 70$ ☐   d) $70 \leqslant M < 80$ ☐

Score   / 3

B

Answer all parts of all questions.

1 Jerry and Kamaljeet are collecting some data on eye colour for a class survey.
Design an observation sheet that they could use. **(3 marks)**

2 Joe and Rhysian are designing a survey to use in the school. One of their questions is shown below:

How much television do you watch?

0–1 hrs	1–2 hrs	2–3 hrs	3–4 hrs

What is the problem with this question? Rewrite the question so that it is improved.

....

.... **(2 marks)**

3 The masses in kg of some students are:

42, 54, 61, 67, 53, 50, 48, 47, 56, 42, 44, 54, 59, 63, 73, 71, 69, 71

Complete the stem-and-leaf diagram below:

4	2 8 7 2 4
5	
6	
7	

stem = 10kg
key 4|2 means 42

(2 marks)

Score   / 7

Answer all parts of the questions.

1 The following data show the weights (*w*) in grams of some packets of sweets.

22	18	4	26	27	21	24
16	8	3	3	12	21	25
14	28	27	10	11	10	14
13	21	6	2	27	18	15

The data are placed into a frequency table. Complete the gaps in the table below.

Weight (*w*) in grams	Frequency
$0 \leqslant w < 5$	4
$5 \leqslant w < 10$	2
.... $\leqslant w < 15$	....
$15 \leqslant w <$	....
$20 \leqslant w < 25$	....
.... $\leqslant w < 30$	6

(3 marks)

2 Some pupils are doing a survey to find out the number of hours their friends spent watching television in a week. They wrote a questionnaire.

a) One question was:

How old are you (in years)?

10 or younger	10 to 15	15 to 20	20 to 25	25 or older

Explain why the labels for the middle three boxes need to be changed.

.... (2 marks)

b) Another question was:

How long do you spend watching television in a week?

don't know		none		a small amount		lots

Jonathan said that the labels need changing.

Write new labels for the boxes. You may change as many boxes as you want.

.... (2 marks)

Score / 7

Total score / 17

How well did you do? ✗ 1–3 Try again 4–8 Getting there 9–12 Good work 13–17 Excellent! ✓

For more help on this topic see KS3 Maths 5–8 Success Guide pages 76–77.

COLLECTING DATA

Handling data

Representing information

A

Answer all parts of all questions.

1 The number of millimetres of rainfall that fell during the first eight days of August are shown on the line graph below:

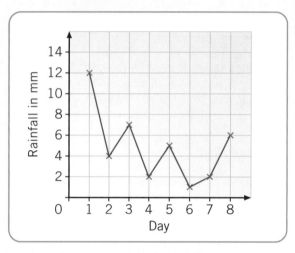

Using the information on the graph, complete the table below.

Day	1	2	3	4	5	6	7	8
Rainfall in mm		4				1		6

(2 marks)

2 The pie chart shows how Erin spends a typical day.

Ⓒ

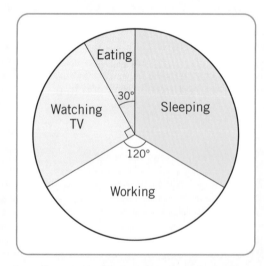

a) Measure the size of the angle for sleeping. ....

b) Work out the number of hours that Erin works. ....

c) For how many hours does Erin watch TV? .... (3 marks)

Score   / 5

 Indicates that a calculator may be used

Answer all parts of the questions.

1   The frequency table and chart show how long (in minutes) 30 people spent queuing at a supermarket checkout.

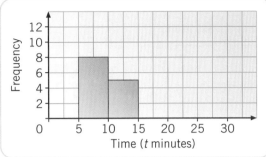

Time (t minutes)	Frequency
$0 \leqslant t < 5$	12
$5 \leqslant t < 10$	....
$10 \leqslant t < 15$	....
$15 \leqslant t < 20$	4
$20 \leqslant t < 25$	1

a) Use the information in the frequency table to complete the frequency diagram.   (2 marks)

b) Use the information in the frequency diagram to complete the frequency table.   (2 marks)

c) Rameen says:

Most people queue for less than 10 minutes

Explain whether he is right.   ....   (1 mark)

2   A teacher asked his class:

C   'What is your favourite type of take-away?'

a) The results are shown in the table below:

Type of take-away	Frequency
Indian	11
Chinese	4
Pizza	8
Fried chicken	1

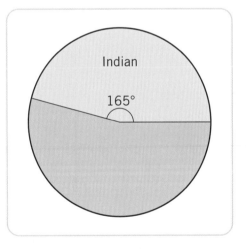

Complete the pie chart to show this information.

Show your working and draw your angles accurately.   (2 marks)

b) What fraction of the students prefer pizza?

   ....   (1 mark)

Score   / 8

Total score   / 13

How well did you do?   ✗ 1–2 Try again   3–5 Getting there   6–9 Good work   10–13 Excellent!   ✓

For more help on this topic see KS3 Maths 5–8 Success Guide pages 78–79.

REPRESENTING INFORMATION   Handling data

75

Scatter diagrams

A

Answer all parts of all questions.

1

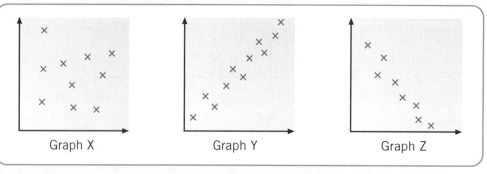

Graph X   Graph Y   Graph Z

Which of the graphs above best illustrates the relationship between:

a) the weight of children and their mark in a maths test?   Graph.................................. (1 mark)

b) the age of a car and its value?   Graph.................................. (1 mark)

c) the age of a car and its mileage?   Graph.................................. (1 mark)

d) the temperature and the sales of ice cream?   Graph.................................. (1 mark)

e) the temperature and the sales of woollen hats?   Graph.................................. (1 mark)

2 **In a survey, the heights of ten girls and their shoe sizes were measured.**

Height in cm	150	157	159	161	158	164	154	152	162	168
Shoe size	3	5	$5\frac{1}{2}$	6	5	$6\frac{1}{2}$	4	$3\frac{1}{2}$	6	7

a) Draw a scatter diagram showing this data.

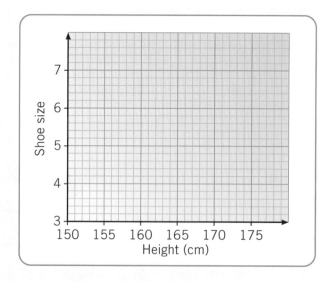

(2 marks)

b) What type of correlation is there between height and shoe size?

.... (1 mark)

c) Draw a line of best fit on your diagram. (1 mark)

d) From your scatter diagram, estimate the height of a girl whose shoe size is $4\frac{1}{2}$.

.... (1 mark)

Score   **/ 10**

Answer all parts of the questions.

1 The following graph shows the sales of some washing powder over a period of 4 months.

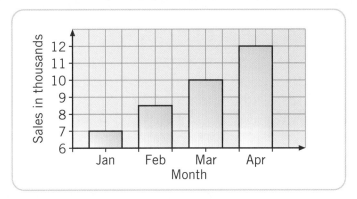

Explain why this graph is misleading. .... **(2 marks)**

2 Some students did a survey of how well they did in a maths test compared to the hours they spent revising and the hours spent watching television. They plotted their results as scatter diagrams.

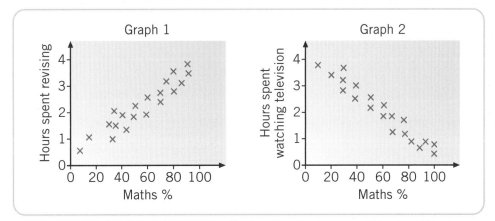

a) What does Graph 1 show about the relationship between the maths percentage and the hours spent revising?

.... **(1 mark)**

b) What does Graph 2 show about the relationship between the maths percentage and the hours spent watching television?

.... **(1 mark)**

c) If a scatter diagram was drawn to show the relationship between the students' maths percentages and their heights, what type of correlation would it have?

.... **(1 mark)**

d) Another student spent 3 hours revising. Using Graph 1, estimate the student's score in the maths test.

.... **(1 mark)**

Score   **/ 6**

Total score   **/ 16**

How well did you do?   ✗ 1–3 **Try again**   4–7 **Getting there**   8–12 **Good work**   13–16 **Excellent!**   ✓

For more help on this topic see KS3 Maths 5–8 Success Guide pages 80–81.

77

Averages 1

Answer all parts of all questions.

1 **Here are some number cards:**

| 2 | 7 | 3 | 6 | 2 | 2 | 1 | 3 | 4 | 2 |

For the number cards, calculate:

a) the mean.. (1 mark)

b) the median .... (1 mark)

c) the mode... (1 mark)

d) the range... (1 mark)

2 **The mean length of 10 tadpoles is 3.2cm. What is the total length of the tadpoles, in centimetres?**

.... (1 mark)

3 **The Chunky Crisp Company claims that, 'On average, a packet of crisps contains 25 crisps.'**

C In order to check the accuracy, a sample of 20 packets was taken and the crisps in each bag were counted.

Crisps	23	24	25	26	27
Number of packets	3	3	12	0	2

a) Find the mean number of crisps per bag in this sample.

.... (2 marks)

b) Explain briefly whether you think the manufacturer is justified in making its claim.

.... (1 mark)

4 **Jessie finds the median of some cards:**

| 7 | 2 | 9 | 4 | 10 |

She says that '9 is the median'. Explain why she is wrong.

.... (2 marks)

Score   / 10

C *Indicates that a calculator may be used*

Answer all parts of the questions.

1 **Ellie has these four number cards:**

| 1 | 9 | 4 | 2 |

The mean is 4.

a) What is the range of the numbers? .... (1 mark)

b) Ellie takes another card:

| 1 | 9 | 4 | 2 | ? |

The mean of the five cards is now 5.

i) What number is on her new card? .... (1 mark)

ii) What is the range of her cards now? .... (1 mark)

iii) What is the mode of her five cards? .... (1 mark)

2 **There are three hidden cards.**

| ? | ? | ? |

The mode of the three numbers is 7.
The mean of the three numbers is 9.

What are the three numbers? .... (3 marks)

3 **In a Sunday league football tournament, the numbers of goals scored are shown in the table below.**

Number of goals scored (x)	0	1	2	3	4	5
Frequency	5	8	9	3	3	2

a) Work out the mean number of goals scored in the tournament.

.... (2 marks)

b) Work out the mode and the median.

mode = ... median = ... (2 marks)

Score / 11

Total score / 21

How well did you do? ✗ 1–4 **Try again** 5–10 **Getting there** 11–16 **Good work** 17–21 **Excellent!** ✓

For more help on this topic see KS3 Maths 5–8 Success Guide pages 82–83.

79

Averages 2

A

Choose just one answer, a, b, c or d.

The following questions are based on the information given in the table below.
The weights of some potatoes:

Weight (W g)	Frequency
$20 \leqslant W < 30$	5
$30 \leqslant W < 40$	6
$40 \leqslant W < 50$	4
$50 \leqslant W < 60$	5

1 How many potatoes weighed less than 50 g? (1 mark)

a) 15 ☐   b) 5 ☐   c) 6 ☐   d) 20 ☐

2 Which of the class intervals is the modal class? (1 mark)

a) $20 \leqslant W < 30$ ☐   b) $30 \leqslant W < 40$ ☐
c) $40 \leqslant W < 50$ ☐   d) $50 \leqslant W < 60$ ☐

3 Which of the class intervals contains the median value? (1 mark)

a) $20 \leqslant W < 30$ ☐   b) $30 \leqslant W < 40$ ☐
c) $40 \leqslant W < 50$ ☐   d) $50 \leqslant W < 60$ ☐

4 Estimate the mean weight of the potatoes. (1 mark)

Ⓒ a) 39g ☐   b) 41g ☐   c) 39.5g ☐   d) 43g ☐

Score   / 4

B

Answer all parts of this question.

1 The masses of some students in a class are measured. The results are given in the table.

Ⓒ

Mass (M kg)	Number of students
$40 \leqslant M < 45$	6
$45 \leqslant M < 50$	5
$50 \leqslant M < 55$	8
$55 \leqslant M < 60$	4
$60 \leqslant M < 65$	2

Explain, with justification, whether the following statements are true or false.

a) The mean mass of the students is 50.7 kg. .... (1 mark)

b) The modal class of the data is $55 \leqslant M < 60$. .... (1 mark)

c) The class interval that contains the median mass of the students is $50 \leqslant M < 55$.

    .... (1 mark)

d) Eleven students had a mass of less than 50 kg. .... (1 mark)

e) There were 20 students in total whose mass was measured. .... (1 mark)

f) The range in masses of the students was 40 kg. .... (1 mark)

Score   / 6

Ⓒ *Indicates that a calculator may be used*

Answer all parts of the questions.

1 The frequency diagram shows the time, in minutes, for a group of students to do a verbal reasoning test.

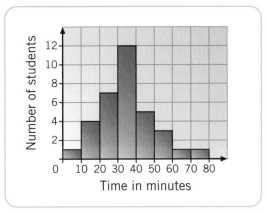

a) Calculate an estimate of the mean time to do the test.

Show your working. .... minutes   (3 marks)

b) Which class interval contains the median time? .... (1 mark)

A different group of students did the same verbal reasoning test. Their results are shown in the table below.

Time in minutes	Frequency (f)	Midpoint (x)	$f \times x$
$0 \leqslant t < 5$	4		
$5 \leqslant t < 10$	18		
$10 \leqslant t < 15$	24		
$15 \leqslant t < 20$	6		
$20 \leqslant t < 25$	2		
$25 \leqslant t < 30$	1		

c) Calculate an estimate for the mean time for this group of students to do the test.

You may use the table to help you... (3 marks)

d) What is the modal class for this data? .... (1 mark)

e) 'On average', which group of students did the test in a shorter time?

Justify your answer. .... (1 mark)

2 Mr Robinson is going to purchase some nails. Two companies give him the following data about the length of their nails:

Company A	Company B
mean = 25 mm range = 8 mm	mean = 25 mm range = 1.3 mm

From which company should he purchase the nails?

Explain your reasoning. .... (2 marks)

Score   **/ 11**

Total score   **/ 21**

How well did you do?   ✗ 1–5 **Try again**   6–10 **Getting there**   11–16 **Good work**   17–21 **Excellent!** ✓

For more help on this topic see KS3 Maths 5–8 Success Guide pages 84–85.

81

AVERAGES 2   Handling data

Cumulative frequency

Answer all parts of all questions.

1 **LEVEL 8** The graph shows the number of mobile phones that a shop sold over a period of 100 days.

Estimate:

a) the median ....

b) the lower quartile ....

c) the upper quartile ....

d) the interquartile range ....

(4 marks)

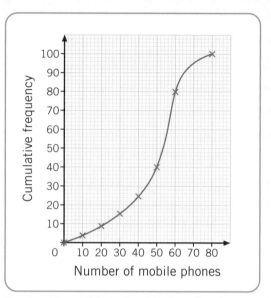

2 **LEVEL 8** The table shows the time in minutes for 83 people's journeys to work.

Time in minutes	Frequency	Cumulative frequency
$0 \leqslant t < 10$	5	
$10 \leqslant t < 20$	20	
$20 \leqslant t < 30$	26	
$30 \leqslant t < 40$	18	
$40 \leqslant t < 50$	10	
$50 \leqslant t < 60$	4	

a) Complete the cumulative frequency column in the table above. (1 mark)

b) Draw the cumulative frequency graph on the axes and grid below.

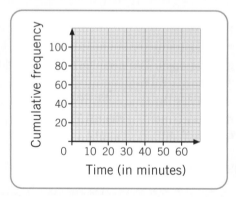

(1 mark)

c) From your graph, find the median time. .... (1 mark)

d) From your graph, find the interquartile range. .... (1 mark)

e) How many people had a journey of more than 45 minutes to work? .... (1 mark)

Score   / 9

B

Answer all parts of the questions.

1 LEVEL 8 **Some students are raising money for charity. The cumulative frequency graph shows the distribution of the amount of money they raised.**

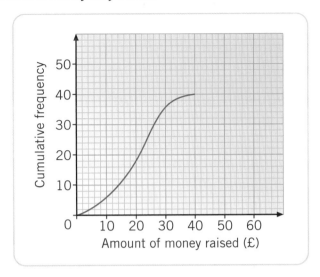

a) Read the graph to estimate the median amount of money raised. £............................ (1 mark)

b) Estimate the percentage of students who raised less than £25. ....% (1 mark)

c) Using the graph, work out the value of the interquartile range.

Interquartile range £.. (1 mark)

Another group of students is also raising money for charity. The table shows the distribution of the amount of money raised by these students.

Money raised (£)	Cumulative frequency
< 10	4
< 20	12
< 30	22
< 40	37
< 50	45
< 60	50

d) Draw a cumulative frequency graph for these data, using the axes above. (2 marks)

e) For the graph you have just drawn:
put a tick (✔) by any statement below which is true,
put a cross (✗) by any statement below which is false.

A: 25 students raised less than £32 each. .... (1 mark)

B: The median amount raised lies between £40 and £50. .... (1 mark)

C: 10% of the students raised less than £5. .... (1 mark)

Score   / 8

Total score   / 17

How well did you do? ✗ 1–3 **Try again** 4–8 **Getting there** 9–12 **Good work** 13–17 **Excellent!** ✔

For more help on this topic see KS3 Maths 5–8 Success Guide pages 86–87.

83

Probability 1

A

Answer all parts of all questions.

1. A bag contains 3 red, 4 blue and 6 green balls. If a ball is chosen at random from the bag, what is the probability of choosing:

   a) a red ball? .... (1 mark)

   b) a green ball? .... (1 mark)

   c) a yellow ball? .... (1 mark)

   d) a blue or a red ball? .... (1 mark)

2. A box of chocolates contains 15 hard centres and 13 soft centres. One chocolate is chosen at random. Work out the probability that it will be:

   a) a hard centre .... (1 mark)

   b) a soft centre .... (1 mark)

   c) a biscuit .... (1 mark)

3. Decide whether each of these statements is true or false.

   a) If the probability that I receive a letter is 72%,
      the probability that I do not receive a letter is 28%. .... (1 mark)

   b) The probability that a flag is not coloured red is 0.26.
      The probability that a flag is coloured red is 0.64. .... (1 mark)

   c) The probability that the 'Go Aheads' win a darts match is $\frac{41}{63}$.
      The probability they do not win a darts match is $\frac{20}{63}$. .... (1 mark)

4. The probability of passing a driving test is 0.7. If 200 people take the driving test today, how many would you expect to pass?

   .... (1 mark)

5. Michelle is a swimmer. The probability of her winning a race is 84%. If she swims in 50 races this season, how many races would you expect her to win?

   .... (1 mark)

Score   / 12

84

Answer all parts of the questions.

1 **The diagram shows two spinners, Spinner X and Spinner Y.**

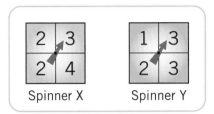

a) Which spinner gives you the better chance of getting a 3?

Explain your choice. .... (1 mark)

b) Which spinner gives you the better chance of obtaining an even number?

Explain your choice. .... (1 mark)

2 **A box contains 50 packets of different flavoured crisps. There are four flavours – cheese, bacon, beef and tomato – in the box. The probability of choosing each flavour from the box is:**

Flavour	Probability
Cheese	0.3
Bacon	0.1
Beef	....
Tomato	0.45

a) Work out the probability of choosing a beef flavoured packet of crisps.

.... (1 mark)

b) Write down the most common flavour of crisps.

.... (1 mark)

c) If a packet of crisps is taken at random from the box, what is the probability
that it is either cheese or bacon flavour?

.... (1 mark)

d) The probability that the box of crisps is damaged during transport is 42%.
What is the probability that the box is not damaged?

.... (1 mark)

e) 200 packets of crisps are chosen at random. Estimate the number of packets of
tomato flavoured crisps in the batch of 200.

.... (1 mark)

Score / 7

Total score / 19

How well did you do?  1–3 **Try again**   4–8 **Getting there**   9–13 **Good work**   14–19 **Excellent!** ✓

For more help on this topic see KS3 Maths 5–8 Success Guide pages 88–89.

85

Probability 2

A

Answer all parts of all questions.

1 **A fair die is thrown 150 times. A 4 comes up 31 times.**

a) What is the relative frequency of getting a 4?

.... (1 mark)

b) If the number of times the die is thrown increases, what would you expect to happen to the relative frequency of obtaining a 4?

.... (1 mark)

2 **A fair die and a fair coin are thrown at the same time.**

a) Complete the table to show the possible outcomes.

		Die					
		1	2	3	4	5	6
Coin	H			H3			
	T				T4		

(1 mark)

b) What is the probability of obtaining a head and a 2?

.... (1 mark)

c) What is the probability of obtaining a tail and an even number?

.... (1 mark)

3 **LEVEL 8** **Mr Smith and Mrs Tate both go to the library every Wednesday. The probability that Mr Smith takes out a fiction book is 0.8, whilst the probability that Mrs Tate takes out a fiction book is 0.4. The events are independent.**

a) Complete the tree diagram by writing in the missing probabilities.

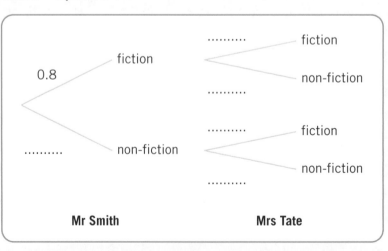

(2 marks)

b) Calculate the probability that both Mr Smith and Mrs Tate take out a fiction book.

.... (1 mark)

c) Calculate the probability that a fiction and a non-fiction book are taken out.

.... (2 marks)

Score   / 10

Answer all parts of the questions.

1 **Hywel and Tracey are playing a game. They each have four cards.**

Hywel's cards are numbered:	1	3	5	7
Tracey's cards are numbered:	1	2	3	4

They each take any one of their own cards. They then multiply the numbers on the two cards.

The table shows all the possible answers:

		Hywel's cards			
	×	1	3	5	7
Tracey's cards	1	1	3	5	7
	2	2	6	10	14
	3	3	9	15	21
	4	4	12	20	28

a) What is the probability that their answer is a multiple of 3?

.... (1 mark)

b) What is the probability that their answer is less than 12?

.... (1 mark)

2 a) Melissa is a keen athlete.

She estimates that the probability she wins each race is 0.3. If she runs 20 races, how many of these races would she expect to win?

.... (1 mark)

b) Simon also runs the same races.

He won 8 of the races he ran. He estimated that the probability that he wins each race is 0.4.

How many races did he run?

.... (1 mark)

c) LEVEL 8 The probability that Imran wins a race is 0.6. Imran runs a race on two consecutive days. What is the probability that he will:

i) win both races?.. (1 mark)

ii) win only one of the races?... (2 marks)

Score   / 7

Total score   / 17

How well did you do?   ✗ 1–3 Try again   4–7 Getting there   8–12 Good work   13–17 Excellent!   ✓

For more help on this topic see KS3 Maths 5–8 Success Guide pages 90–91.

87

PROBABILITY 2   Handling data

Notes